Cosmic Impressions

COSMIC
IMPRESSIONS

Traces of God in the Laws of Nature

Walter Thirring

Translated by Margaret A. Schellenberg

TEMPLETON FOUNDATION PRESS
PHILADELPHIA AND LONDON

Templeton Foundation Press
300 Conshohocken State Road, Suite 670
West Conshohocken, PA 19428
www.templetonpress.org

German edition © 2004 by Molden Verlag GmbH, Wien
English translation © 2007 by Templeton Foundation Press

Templeton Foundation Press helps intellectual leaders and others learn about science research on aspects of realities, invisible and intangible. Spiritual realities include unlimited love, accelerating creativity, worship, and the benefits of purpose in persons and in the cosmos.

Typeset and designed by Gopa & Ted2, Inc.

Library of Congress Cataloging-in-Publication Data
Thirring, Walter E., 1927-
   [Kosmische Impressionen. English]
   Cosmic impressions : Traces of God in the laws of nature /
Walter Thirring.
      p. cm.
   Includes bibliographical references and index.
   ISBN 978-1-59947-115-0 (pbk. : alk. paper) 1. Religion and
science. I. Title.
   BL240.3.T4513 2007
   215—dc22
                                    2006037633

Printed in the United States of America

07 08 09 10 11 12   10 9 8 7 6 5 4 3 2 1

Dedicated to my wife for the long journey we've shared

Contents

Foreword

IT WAS with pleasure that I accepted the suggestion from the publisher to say a few words about the timeliness of the chosen topic for this book of Professor Thirring, the former director of Vienna University's Institute for Theoretical Physics, which deals with the traces of God in the creation.

The strained history of the relationship between a scientific and a religious view of the world seems to have been alleviated in recent years after a long period of opposition since the Enlightenment. The religious pluralism in our world, which is becoming ever more unified, makes it easier to see that religion, all religions since time began, have sought an answer to the unsolved mysteries of human existence: Where did I come from? Where am I going? What is the meaning of life? What is the last secret of our existence from whence we come, and where are we headed?

Viktor Frankl, who started out as a follower of Sigmund Freud, rephrased the question of the meaning of life for our time with his famous logotherapy. The answer is not identical to the answer of whether God exists, but leads us in this direction, as Frankl himself thought. A history of the religions of humanity, a comparative study of religions with a wealth of new insights, definitively demonstrated that there has never been a people or a tribe in the history of humanity without a certain kind of religious questioning and ritual. This is also an indication that today the deterministic worldview, which seemed indisputable due to evolution, is losing its powers of persuasion. As the Nobel Prize winner and director of the European Organization for Nuclear Research (CERN) Carlo Rubbia stated in an interview with the Swiss newspaper *Neue Züricher Zeitung* in March 1993, "As a research scientist I'm very impressed by the order and beauty I find in the universe as well as within material things. And as an observer of nature I can't dismiss the idea that there was a higher order existing before the things. . . . This stipulates an intelligence on a higher level, beyond the existence of the universe itself."

The former contradiction between religion and science doesn't appear to exist still in the form it had originally taken. New connections and orientations are being discovered. This is clearly expressed in the book *Cosmos, Bios, Theos*, where a series of internationally renowned experts from the fields of astronomy, physics, and mathematics, including twenty Nobel Prize winners, were asked about their personal relationship to religion and faith. The answers were published, and they included the answer of the author of our present book. He believes we can see the hand of God mysteriously guiding through the theory of evolution, previously thought to be indisputable. At any rate, this colorful series of personal answers and short commentaries proves that new insights and understanding leading to a new orientation are coming into being. Our present book fully shares this concern.

I am glad that Professor Thirring has provided a new contribution to enable the dialogue between religious, Christian faith and science to continue in this way and develop for the good of humanity. Similarly, over sixty years ago when the dropping of the nuclear bomb terrified the world, Albert Einstein warned that the atom bomb had changed everything in the world, except the way we think. This has to change if humanity wants to survive.

Cardinal Franz König

Preface

The ideas in this book originate from several different sources.

I WAS FORTUNATE enough not only to have met the great architects of the scientific worldview such as Einstein, Heisenberg, Schrödinger, and Pauli, but also to have been able to get to know their ideas directly in discussions. I would like to pass these impressions on, not necessarily to further glorify these thinkers, but to present them in the way they live in my memories, with all of their fascination and idiosyncrasies.

I had always been in awe of the miraculous blueprint of the cosmos. However, it's written in the language of mathematics and not accessible to the majority of people. Through simple estimates, a rough outline can be divined, and even this offers a majestic panorama. I have tried to convey these reflections in lectures and have always been urged to present them to a larger audience. My goal in this book is to paint this picture without overwhelming the majority of readers, even at the risk of boring some with too many details. When my ruminations nonetheless became somewhat heavy-going or when more formulas were necessary, I've removed these passages from the text and collected them in appendixes. Of course, the reader can ignore these and then simply accept the facts and figures given. I don't want to overload the text with exact figures. When a reader repeats the calculations, he or she must keep in mind that I have rounded up or down to the next power of ten. In addition, symbols and terms are also explained at the end of the book. It is impossible to speak about the results of quantum theory logically in everyday language. It is not the reader's fault if these arguments are sometimes impossible to follow.

In the end, reflections on the creation of the universe lead to reflections about the Creator, which may surprise some readers to hear. It's also common to hear the opinion that science leads to atheism. This is an opinion that I can't share; I even find it somewhat absurd. When we are moved by a fantastic building, a cathedral, or a mosque and have finally realized what is behind the glorious proportions, who would then say, "Now we don't need the architect anymore. There might not even be one, this could all just be the random product of circumstance"?

What conclusions can we draw beyond the feeling of awe? I can't follow wider-reaching theological speculations; sometimes I find them to be blasphemous. How dare we presume to capture God in our web of logic? I personally am moved just by simple statements like the parables or the seven petitions in the Lord's Prayer in the New Testament. That's why the reader will not find proof of God's existence in this book. Every argument for the existence of God requires certain logical rules. Who would even dare to dictate to God the rules by which he would have to manifest himself? Therefore, I don't want to say that the order of the universe proves the existence of God. I prefer the wording of the psalm of David, so beautifully set to music by Ludwig van Beethoven:

The heavens extol the glory of God.

It should hardly be necessary to mention that all of my observations are made from the point of view of modern science. It can always be said that we will someday have a different and better understanding. This may be true, but is irrelevant for now. We have to take into account that which is on offer, and that in itself is already overwhelming.

In the last decades, new worlds have been unveiled that our great teachers wouldn't even have dreamed of. The panorama of cosmic evolution now enables deep insights into the blueprint of creation. I would like to discuss these, and I hope to convey my enthusiasm to the reader in this book. Even though everything happens according to certain basic laws, in each developmental stage of evolution random events have played a decisive role and conducted things so that the cosmos was ultimately crowned by the human spirit. Human beings recognize the blueprints, can understand the language of the Creator and can raise themselves to his image. These realizations do not make the sciences the enemy of religion, but glorify the book of Genesis in the Bible.

Cosmic Impressions

How the World Came into Being

GENESIS

Which main ideas of the Bible are substantiated
by modern knowledge?

In the beginning God created the heavens and the earth.
The earth was without form and void, and darkness was
upon the face of the deep; and the Spirit of God was moving
over the face of the waters.
And God said, "Let there be light," and there was light.
(GENESIS 1:1–3)

THESE LINES have formed Western thought and inspired artists and musicians to glorious achievements. We have always been interested in the question of where it all comes from, and different peoples have developed their own mythologies. Until recently, there were no empirical data available, so that one could let one's imagination run wild. The universe became home to innumerable gods, heroes, and all kinds of creatures in the various perceptions of the world. Some things were hard to explain, and some ideas were very clear and concrete, such as Eastern mythologies where an elephant stands on top of a turtle (or the other way around). The significance and amount of truth in such statements have been subject to much speculation. A scientist who has been trained to accept only what can be proven would naturally have problems here. Wolfgang Pauli used to say about statements that could neither be proven nor refuted, "That's not even wrong, but complete nonsense."

It therefore didn't take long for Genesis to come under criticism, and some contradictions between what it says and scientific knowledge even led to doubts about the Bible itself. Before we start discussing scientific corrections and improvements, we must first put the question itself in the correct light.

Everything can be described at several levels. We can either concentrate on the essentials and touch on some details inexactly or even incorrectly, or list particulars that will only be important at another level and in the process lose sight of the main point. This is true in every field, including science. When chemists call water Aich-Two-Oh or write H_2O, then they are pointing out the main characteristic of water, namely, that it is formed by molecules from two hydrogen atoms and one oxygen atom. There are still many details that this simple structure does not explain, like how the atoms join together to form a molecule, what the lines in the structural formula mean, the angle of the HOH atoms, and so on. A physicist would even say that the choice of words was inappropriate. For a precise description, you need to solve the Schrödinger equation for ten electrons in the energy field of two protons (hydrogen nuclei) and an oxygen nucleus. Unfortunately, this exceeds the capacity of even our present-day computers. All these subtleties don't even touch upon what the engineer wants to know—that is, how hydrogen (H_2) and oxygen (O_2) should be mixed so that the gases fully burn up during an exothermic reaction. The engineer just needs to know that the substances need to be combined with the ratio of the number of molecules equaling 2:1.

The significance of Genesis can be seen in light of what has just been said. Written at the time of the Babylonian imprisonment of the Jews, it has a very clear message. It wants to tell the Jews that their God created the world and has always been with them ever since. All of these scientific details were unavailable then, or else they would have diverted attention away from the main point. The creation of the world by one God is a significant step in human thought, as it leads to a strict monotheism. This doesn't only say that every people should have just one God, but that there is only one true God in the whole world. Unfortunately, not only the Creation but also the surroundings were taken literally, which corresponded to the ideas of that time. This misunderstanding has led to great harm in the development of Western thought. Some independent thinkers faced persecution or were even burned at the stake, and religion lost some of its credibility. Despite everything, it's remarkable that there are many parallels between what's in Genesis and what's in the modern, scientific view of the world. Therefore, at the end of this chapter, I want to present an accompaniment from the world of physics to a biblical text, which would be difficult for me to do with other mythologies. In some sense the message in Genesis about the omnipotence of Yahweh was further confirmed by the fact that the laws of nature are not only valid between Babylon and Jerusalem, but also, as we know today, over billions of light years away. The same laws apply even in areas so far away from each other that they aren't able to communicate with one another without breaking the speed limit of light. Yahweh's power is much, much greater than originally thought.

SIZE AND AGE OF THE WORLD

In the universe, ten billion years is the essential time,
ten billion light years the essential length.

The architecture of the cosmos is determined by dizzyingly large numbers. First, we will need to see how we can deal with them. The best thing is for us to write them using exponents—that is, 10^6 for 1,000,000 or 1 million—the exponent always showing the number of zeros. We'll save more exact instructions on how to use them for appendix 2 so as not to interrupt the text flow. For now, we want just to simplify word monsters like a million million million to 10^{18}. In multiplication, the exponents are added together, and so we can reduce the necessary math to simple arithmetic.

To see the usefulness of using exponents, we'll estimate the number of atoms in a cubic meter (m^3) as a warm-up exercise. For that, we will express the size of different objects as exponents of a meter (m). An atom is about 10^{-10} m large, so that 10^{10} atoms fit on a meter and $10^{10} \times 10^{10} \times 10^{10} = 10^{30}$ atoms in a m^3. Now the discerning reader will object that it can't possibly work out so neatly. To avoid such trivial objections, we will therefore omit the equal sign (=) when we can and use the following symbols instead:

a ~ b   a is about equal to b,

a > b   a is significantly greater than b,

a < b   a is significantly less than b.

Of course a pedant could ask what exactly "about equal" or "significantly greater" means. That is not the important thing for us, however, and we will boldly write that the number of atoms in a m^3 of solid matter is ~10^{30}.

After we've overcome our reservations about large numbers, we can treat ourselves to using the same system of units all the time, so we won't always have to convert. It doesn't actually matter which system of unit we choose; we can bow to modern pressure and measure length in meters (m), mass in kilograms (kg), and time in seconds (s). We will stick to these as much as possible (please direct eventual complaints to the international measurement commission), but will fall back on popular sizes like the kilometer from time to time.

Hardly any other number thought to have been discovered in the Bible has led to so many faulty perceptions as that of six thousand years for the age of the world. The disapproving stance taken by the church toward some scientific developments, such as Darwin's theory of evolution, can be traced back to this. There's no way a worm can develop into a human being in such a short period of time. If we interpret the parts of Genesis as earth days, then we have to come up with these six thousand years. In the Gospels of Matthew and Luke, Jesus' (or better Joseph's) family tree is traced back to Abraham, and in Genesis from Abraham to Adam. When we then add in the week of Genesis, we still won't come up with more than a few thousand years, even taking into account the Methusalean age people seemed to have reached back then. This six-thousand-year time span really did play an important role in human development. At that time, high cultures flourished all over the earth. This must have been related to the fact that navigation skills had been developed to allow access to large areas via extensive river systems. In this way it was possible to make laws valid for huge empires, which enabled the regulated coexistence of large nations. Six thousand years ago might have been when laws were created, but it wasn't the first days for the laws of nature.

The idea that everything began six thousand years ago proved to be extremely pervasive; it was still being mentioned in a letter from Richard Wagner. This number was then revised by Lord Kelvin at the end of the nineteenth century by a factor of 1,000. He believed to be able to show that the sun was not older than a few million years; otherwise, sunlight would have consumed the gravitational energy of the sun, and it would be nothing more than a cold, lifeless mass. I don't remember how Lord Kelvin's argument went, but I know how it should have gone. I hope that the reader has gotten used to using exponents because I would like to invite everyone to reconstruct this argument in appendix 3. The sizes gained from that will be important for everything coming up.

Lord Kelvin's result that the sun was only a few million years old was, scientifically speaking, the death knoll for Darwin's theory of evolution, because a worm still can't become a human being in a million years. Only the transformation from the higher apes to human beings could have been accomplished in such a period. There didn't seem to be a way out. Things would have stayed this way if Sir Arthur Eddington in 1920 and Robert Atkinson and Fritz Houtermans in 1929 hadn't had the idea that nuclear energy could prolong the life of a star by another factor of 1,000 as a stellar thermonuclear reactor. This energy was unknown in Kelvin's day. For seven years, I was the assistant to and coworker of Fritz Houtermans, the Vienna-born physicist who was without a doubt an interesting man. I personally owe him a lot because he gave me a job when I was an unemployed physicist and gave me a chance to have a career in science. I was also engaged to be married at that time, and thanks to the job with him in Berne, my fiancée and I could get married; Houtermans was our witness. Therefore, I'm not completely objective when describing him, but I don't want to digress and will speak more of him later. I would only add that the night after having had this paradigm-building idea when he was twenty-six, he was out for a walk with his girlfriend. She said, "Look at how beautifully the stars are shining," and he replied, "Since this morning I know why they're still shining."

Nuclear physics was still underdeveloped, or more accurately, nonexistent in Atkinson's and Houtermans' day. Their work was premature and got the details wrong. The fact that this kind of an energy source can tap into a whole other magnitude becomes clear on examination of the simplest facts. Lord Kelvin observed that the gravitational energy is as great as that of a body crashing into the sun due to gravity. In the next chapter we will see that during a decay the a particles of a nucleus (the nucleus of a helium atom) whir around at the unbelievable speed of 30,000 km/s. In appendix 3 we have determined that the speed of falling onto the sun from a great distance will be 300 km/s at the surface, 10^2 less than the speed of the a particles.

The kinetic energy increases with the square of the speed; therefore, the energy of the particles in a nucleus is 10^4 times greater than the gravitational energy of the particles in the sun. We gain just as much through nuclear energy for the energy reserves of the sun and can enjoy sunlight for billions of years without the sun running out of breath.

The discerning reader will realize that this argument talks about how old the sun can be, but nothing about when its creation took place. The aforementioned a-decay shows that at least the earth has existed for billions of years. There are elements like uranium whose life span lies in this magnitude and then they decay. We only need to look at how much decay product has gathered in deposits of uranium. From that, it's possible to determine the time since the decay product has remained with the uranium—that is, since there have been regular conditions on the earth's surface. These reflections yield an age of 4.5×10^9 years for the earth.

There is evidence that supports the idea of the entire universe being 10^{10} years old—that is, only two or three times older than the earth. Whether it's twelve or fifteen billion years is still being disputed, but this is irrelevant for our purposes. One part of Genesis has been confirmed, and that is that the world began at a particular time.

After we have come up with a cosmic age of 10^{10} years, then estimating the size of the visible universe is easily possible. It began as a ball of fire expanding almost at the speed of light. Its size is therefore the distance that light travels in 10^{10} years, or 10^{10} light years.

How Do We Know That the Universe Is Expanding?

We know this thanks to an effect detected by the Salzburg physicist Christian Doppler more than 150 years ago.

When you hear a train's whistle blow, its tone is higher when the train is approaching and lower when it is receding. The situation is analogous for sources of light. If they are approaching the observer, they appear to be bluer; when they are receding, they become redder. When the light of known elements in an astronomical object appears redder than normal, then we know that it is moving away from us and that the size of this "red shift" tells us its speed. In this way, although not with the naked eye, we can directly observe that the galaxies are moving like fragments of a great explosion in the sky. The farther away they are, the weaker they shine; the weaker they shine, the faster they are fleeing.

The age of the universe is gained by multiplying the universe's age of $10^{17.5}$ seconds by the speed of expansion, i.e., $10^{17.5}$ s × the speed of light = 10^{26} m. This enormous figure exceeds our imagination and also doesn't tell us much at first. The size of the universe becomes more comprehensible when we compare its structure with that of matter. Let's imagine the universe as a gas by shrinking the planetary systems to the size of atoms. The stars would be the atomic nuclei.

The nearest star is over a light year away from the sun: ~10^{13} km, i.e., 10^4 times the size of our solar system. Under normal conditions, the distance between atoms in a gas is around 10 times their diameter. In our thought experiment, the firmament appears as a strongly diluted gas with a distance between the "atoms" 10^4 times the atomic radius rather than just 10 times.

Stars clump together to form the next largest structure, galaxies, which correspond only to tiny droplets in the analogy to gas. A typical galaxy has the approximate volume of:

$$\text{height} \times \text{width} \times \text{length} = 10^4 \times 10^5 \times 10^5 \text{ (light years)}^3$$

and with an average distance between stars of ten light years contains:

$$(10^4/10) \times (10^5/10) \times (10^5/10) = 10^{3+4+4} = 10^{11} \text{ stars.}$$

On the other hand, fog droplets contain many more atoms, perhaps 10^{18}.

The distance between galaxies is roughly 100 × their height ~10^6 light years. Because the universe is 10^{10} light years large, there would be enough room for 10^4 galaxies in one direction; in all three together $10^4 \times 10^4 \times 10^4 = 10^{12}$ galaxies. Improved telescopes have shown that the universe largely has a structure similar to a sponge with lots of empty spaces and that it actually contains only 10^{11} galaxies, ~10^{22} stars. The universe appears homogenous in an even broader overview.

With that we've come to the end, and in our mini-model the visible universe is a ball 10 km in diameter. It's filled with 10^{22} atoms, just a spoon tip full of matter, because a spoonful of matter contains 10^{24} atoms. The galaxies in this miniature model float as 1 cm large droplets about 1 m

Figure 1.1: Views of the universe in different magnifications. Upper left: a computer model of part of randomly distributed galaxies. Upper right: the actual distribution of galaxies is somewhat clumpier because of gravitation. Lower left: part of a photo from the Hubble telescope of the farthest corner of the universe. Lower right: spiral galaxy NGC 4603, taken with the Hubble telescope. The magnification per photo increases by a factor of around 100.

away from each other. Our living conditions are certainly neat and tidy, aren't they?

WHERE DOES ALL THE ENERGY COME FROM?

The creation of the universe seems to break the limits of science and yet can co-exist with the laws of nature.

The Russian émigré George Gamow was one of the greatest visionaries of the twentieth century. He knew that his visions were just speculations at the time, and he didn't pronounce them to be the ultimate truth, but as ideas not to be taken quite seriously. Maybe this is why he wasn't taken seriously at first, but history has proved him to have been mostly right. We have already crossed paths with him unwittingly. The realization that solar energy is originally nuclear energy is based on one of his conclusions. I just want to briefly sketch out what this means. It was the achievement of Atkinson and Houtermans to recognize that, according to quantum theory, the fire in the core of a star can be fed by nuclear reactions, although the traditional idea (from now on called "classic") says that this is impossible. Atomic nuclei all have positive electric charges, and the identical charges repel each other. It's only at extremely small distances (10^{-15} m) that the attraction of nuclear forces plays a role and the nuclei can cling together. Until then they must overcome the huge barrier of electric repulsion, and the energy of the measly million degrees that we've come up with in appendix 3 is no match for that. We've already seen how another whole magnitude of energy exists in the nucleus. According to quantum theory, however, the miracle can sometimes happen that a nucleus can slip through this barrier. Later we will see that particles can even come to light from the condition of nonexistence, the so-called virtual particles. Normally, miracles don't happen too often; using quantum theory Gamow was the first to calculate just how frequent they do. This tiny probability is now called the Gamow factor. We will also meet this Gamow factor at a decay because the a particle first must overcome the attraction of nuclear energy before it can be spun out electrically. According to the classic view, this would be impossible; with the quantum theory it can happen after countless attempts. Just how often this happens is also revealed by the Gamow factor, but by an a decay it's working from the other (inner) side. In a star it tells us how often nuclei, despite electric repulsion, manage to come together and melt together to form a larger nucleus while producing a large amount of energy. This is, according to Atkinson and Houtermans, the most productive energy source for stars. We will encounter Gamow later on as he lays the stars out for their final rest. Here he was their energy donor, and he showed us the rest energy in atomic nuclei.

Let's now discuss our understanding of the creation of the universe from today's perspective. Whatever happened in the first three minutes cannot be directly observed and remains subject to speculation. Therefore, the version of events offered in this chapter is not as certain as those offered in the others. Nonetheless, we want to venture all the way back to the big bang. We're just interested in working out what modern science seems capable of explaining. These ideas might someday be revised again. However, the following should remain true: ever since the very beginning, randomness and necessity have been working together to create the world to be the way it is today.

Gamow also suggested the big bang theory, according to which the universe came into being after an enormous explosion and all that we can still see are the bits of it flying away. Many physicists didn't like this idea as it sounded like some sort of pumped-up muscular God who magically conjured up immense energy out of nothing.

They were pleased when an alternative theory appeared on the scene, the so-called steady state theory, according to which, the universe has neither a beginning nor an end, but just continues lolloping around. Gamow's theory was greeted enthusiastically by the church; Pius XII practically canonized it. Vatican dignitaries looked rather askance at him for this; there really wasn't much evidence for the big bang at that time. Since then, though, it has turned out that Pius XII had a sound intuition for physics, and today the evidence for the big bang is so overwhelming that hardly anyone has any doubts that it somehow took place. Three pillars support the big bang theory:

1. The aforementioned red-shift effect. Its existence has been further demonstrated through an immense increase in and improvement of observation material.

2. Due to the expansion of the universe, the flash of the big bang has degenerated to a weak form of radiation. We can't see this with the naked eye, as it now only has its maximum intensity in microwaves. Today radio astronomers are not only able to identify this flash, but can also measure the spectrum of radiation to the highest degree of accuracy. It turns out that its light has cooled down to $3°$ K (absolute degrees), just as would be expected with the expansion. It has a spectrum like the light in a thermal balance, which must have dominated in the forges of the universe in the beginning.

3. Even if only the simplest parts of matter—electrons and protons—were created shortly after the big bang, the temperatures must have been so great that nuclear reactions took place. The first obstacle for the development of the elements appeared: two protons do not stay together, no matter how strongly they're blasted together. Luckily other particles jump

in at this point. A proton (P) can virtually, meaning by taking out an energy loan, transform into a neutron (N), a positron (e^+), and a neutrino (v). If another proton happens to be around, the neutron can stick to it to build a deuteron D (atomic nucleus: P + N, heavy hydrogen), and the rest escapes according to the reaction: $P + P \rightarrow D + e^+ + v$. If protons and neutrons stay together, they can pay back the energy bank with the bond energy created, and everything evens out. We still haven't gotten past hydrogen; D is an isotope of hydrogen, but two deuterons can built another proton and tritium T (atomic nucleus: P + 2N, very heavy hydrogen), which is what happens in the reaction D + D = T + P. This is possible because the tritium nucleus contains one proton and two neutrons. The reaction in diagram form is: (P + N) + (P + N) becomes (P + 2N) + P, with the number of protons and neutrons staying the same. At this point a helium nucleus (a particle) can finally be made: D + T = a (He) + N. The number of protons and neutrons remains the same in this reaction, too (P + N) + (P + 2N) $\rightarrow$ (2P + 2N) + N, so that it can take place without further ado.

You would think we could work through the entire periodic table of elements in this way, but there are two gaps in this system. On the one hand, a proton will not stick to a helium nucleus. On the other, two a particles will not stay together; the beryllium nucleus ^8Be = (4P + 4N) has an extremely short life before it decays. Another a particle needs to appear to form the carbon nucleus ^{12}C = $3a$. However, there wasn't that much time at the big bang; everything flew apart too quickly. The heavy elements could only be hatched in stars, where everything is held together for billions of years. If Gamow's big bang theory is true, then the universe, as complex calculations indicate, should contain matter that is about three-quarters hydrogen and one-quarter helium, sweetened with heavy elements from the star's kitchens. This is indeed the composition found.

The three pillars of the big bang theory are completely independent from one another, but combined they form a trustworthy, solid fundament. However, the question as to where all of this energy came from remains unanswered. This puzzle will be solved by Einstein's theory of gravitation.

Energy is the cosmic currency. Various forms of energy can be changed into one another, but the sum total must be correct (except for temporary fudging in quantum mechanics). Visible energy contained in matter and nonvisible in gravity together form the total energy. We've already encountered the first kind with the kinetic energy of a body; it increases with the square of its speed. However, even when a body is at rest, it contains an incredible amount of energy. This is the so-called rest energy, which is given in Einstein's famous formula made infamous by the atomic bomb:

$$E = mc^2 \ (c = \text{the speed of light})$$

The dormant energy in a body at rest is twice as great as its kinetic energy would be if it moved at the speed of light.

Before we calculate the total energy of the universe, I'd like to take a break and talk a little about Albert Einstein, as he will dominate this chapter.

Figure 1.2: Einstein gave me this photo in 1953.

Einstein was without a doubt the greatest physicist of the twentieth century; in all of history, perhaps only Isaac Newton comes close. Einstein enriched physics in many ways; he became popular because of the unfortunately named theory of relativity. It consists of two parts: the special theory of relativity as the logical completion of electrodynamics and the general theory of relativity as an improvement to the Newtonian gravitational theory with entirely new concepts.

Einstein's gravitational theory is based on the realization that gravity forms space and time. At first the theory of relativity excited some philosophical thinkers, but for practical thinkers its effects were too small to be significant. Times have changed, and now not only economic interests but also the personal safety of thousands of people are dependent on it. The GPS (global positioning system) has the fantastic accuracy of four meters only because of corrections made with the theory of relativity. Without it, the margin of error would soon become as much as 11 km, and GPS would have been useless.

It was my privilege to get to know Einstein personally and to discuss several questions in physics with him. This came about because of the work that I will describe in chapter 2, pages 44–48, which generated an invitation to the Institute for Advanced Studies in Princeton for a year, where Einstein worked. Although he was already an old man at that time, he was still open to new people and had an impish sense of humor. This became apparent as our first conversation took an unexpected turn. When he asked me where I was from, I answered that I was currently working as an assistant at the University of Berne. His face became transformed, and he said that he had had a good time in Berne when he was young and had learned a lot about physics there. I was surprised by the last remark, because I knew that he published three articles in Berne in 1905, one of which later won him the Nobel Prize while the others could have. However, there wasn't anyone around who came close to his caliber, so whom could he have learned something from? Einstein told me he used to like to go watch feeding time at the bear caves in the old town. He observed that the bears usually walked with their mouths to the ground and would only find what was in front of their noses. Sometimes one would get up onto its hind legs and could see from its perch the really good treats. This reminded him of the physicists who were usually bent over their calculations and only saw the next equation that was in front of their noses. The most important discoveries are made only when you can see the greater connections. He never showed particular interest in isolated phenomena, like the details of how a helium atom functioned. He was concerned with the fundamentals; he wanted to "get wise to him." This conversation inspired me and the title of this book.

I'll talk more about my physics discussions with Einstein later on, but for now I'd like to get back to the energy of the universe. As we've already mentioned, visible energy is the sum of two parts: the rest energy mc^2 and kinetic energy. The nonluminous part of total energy, the gravitational energy, is the product of its mass and a gravitational potential that varies from place to place and is nothing more than a mental construct. If a body is pulled by the gravity of a large mass, then it falls toward it with a speed determined by the conservation of the mechanical energy (total energy less rest energy). If the path of a test body begins at the speed of zero in a place without gravitational potential, then its mechanical energy will also remain zero. If this body begins to fall, then the kinetic energy will become positive, which means that the gravitational energy must become negative to maintain the zero-sum game. Both sums contain the mass of the test body as a factor, and therefore it can be concluded from the energy balance that its speed is not dependent on its mass. Galileo had already demonstrated this in his famous experiment at the Leaning Tower of Pisa. Bodies with different masses fall at the same speed as long as they are sufficiently heavy and not blown around in the air. In the experiment from the Leaning Tower, the gravitational potential was spawned by the earth. In general, the gravitation potential generated by an object is proportional to its mass and inversely proportional to the distance to the center of gravity. The important thing to keep in mind is that this amount is negative.

Still within the framework of these simple reflections, let's let our test body fall into the gravitational hole of the universe. How fast will it be then? For this we need the mass M_U and the radius R_U of the universe. It's true that we still haven't weighed the universe in kilograms, but that doesn't matter. We know that it contains 10^{22} stars, on average perhaps about as heavy as the sun, and that its radius R_U is 10^{10} light years. This gives us all the data we need to do our homework. We just need to compare it to the crash into the sun. This simple calculation is laid out in appendix 4; we'll come straight to the result here. When falling into the universe, the speed v_U compares with the speed v_s when falling onto the sun as follows: $v_U \sim v_s \times 10^{2.5}$. As v_s was $10^{2.5}$ km/s, v_U is approximately 10^5 km/s. We see that we're getting close to the speed of light $c = 10^{5.5}$ km/s when falling into the universe. This doesn't just make us a bit uneasy; the kinetic energy would then have to be as large as the rest energy, and the gravitation energy is always its negative, to maintain the zero sum. The gravitational hole of the universe must be as deep as the rest energy of the particles. Although the gravitational force is usually small, the gravitational energy will be so enormous that the *creatio ex nihilo* (creation out of nothing) at least doesn't impinge on the conservation of energy. Our reflections have shown that the gravitation energy—GM_Um/R_U in the well is exactly the negative of the rest energy mc^2 of our body, and thus the sum energy is zero. The test body is nothing special: all of us, even the entire universe has the sum energy of zero. Thus, when God created the universe, he didn't need to plunge himself into an energy debt. He organized everything with a balanced budget.

Before we get too carried away, we need to deal with our sense of unease at our approach toward the speed of light. Naïve considerations are no longer valid under such extreme conditions, and Einstein's gravitational theory has to hold out. I should also emphasize that the philosophical objections to *creatio ex nihilo* are barely touched on in all this. Nothingness in the philosophical sense doesn't even exist in modern physics; the fields of elementary particles are also present in a vacuum. Before we get into heavier fare, I'd like to tell the story of the pioneer who discovered what happens to Einstein's theory in this situation.

At the beginning of the twentieth century, there was an astronomer named Karl Schwarzschild. For years, he was the private astronomer to a wealthy patron in Vienna before he became a professor in Germany. In World War I, he volunteered for the front, first going to the Western and then to the Eastern front. There, before his death, the events took place that would earn his name a permanent place in the history of science. In 1916, Einstein published his new gravitational theory, and Schwarzschild got hold of a copy of it on the front. His genius flourished once more. He immediately grasped the meaning of Einstein's work, which went beyond most contemporary physicists' understanding at that time. He was even able to make an exact solution to these incredibly complex equations, something even Einstein couldn't do. Then death ripped the pen from his hand.

Since then generations of physicists have made careers out of trying to solve the enigma of the so-called Schwarzschild solution. This solution to Einstein's equations digs a gravitational hole deeper than the particle's rest energy and contains the entire paradigm change of the general theory of relativity. (The trendy-sounding "paradigm change" just means that the new terms began to seep through to scientists' brains.) The Schwarzschild solution describes an even more extreme condition than we found in the situation outlined for the universe. What happens when a gravitational hole becomes so deep that a particle falling into it moves faster than the speed of light? It took almost half a century before it was noticed that this was something like a mouse trap. Nothing can get out of such a hole; you'd have to run out of it faster than the speed of light, which would be a problem. John Archibald Wheeler came up with the term "black hole" to describe this part of the Schwarzschild solution, which today dominates the scientific popular press.

In Schwarzschild's calculations, everything becomes infinite at a distance from the mass M, where gravitational energy equals the rest energy. This seemed to be a place of death; today this distance is called the Schwarzschild radius. It was only over the course of time that people realized that this kind of infinity was based on the coordinates used. The Schwarzschild radius just means that beyond it is the point of no return. Using better coordinates, we can follow the Schwarzschild formula toward the center, and then we find a place where magical things happen, like in a fairy tale. It's possible to meet people from other worlds whose existence we know nothing about. Unfortunately, this surprise doesn't last long, as then you'd be crushed by the extreme gravity. This shows that, under certain conditions, our known laws of nature demand the existence of other universes that we can't see.

Although they have long been misunderstood and subject to doubt, modern observations leave no room to doubt the existence of black holes. The most terrifying phenomena in the universe, they are the ultimate hangman and gravedigger. They devour complete stars to return them to the underworld of nonexistence from whence they came that first day.

Getting back to our universe, according to what we've learned so far, it's something like a giant black hole. With its overpowering gravitation, it binds everything to itself and tries to squash it. However, we're still a long way away from the point where everything collapses. The black holes found by astronomers are purely local phenomena. Today, there's evidence that the complete collapse of the entire universe will not take place, that its expansion seems to be accelerating. As we've seen, which energy dominates and how the universe will further develop is a touch-and-go situation. How the situation is evaluated is mostly a question of personal taste. My friend Freeman Dyson once said that he wouldn't want to think about a closed universe, because then he'd get claustrophobic. On the other hand, my wife would feel a sense of security in a closed universe.

To end this section I'd like to return to my conversation with Einstein.

Figure 1.3: I took this photo of Einstein in 1954, one year before he died.

I was working on the question of whether, in quantum theory, with a gravitational hole deeper than mc^2, particles begin to seep out of the underworld and it would be possible to tap into an endless reservoir of particles. Einstein's reaction was a disappointment because he didn't believe in quantum theory. First, he didn't want to accept what I was saying and thought I was talking about cooking up new elements in stars. When I replied that this was absolutely not the case, but rather I was talking about particles coming into being from nothing, he was shocked. He had grown used to some of the paradoxes of quantum theory, but that was too radical for him. I then reined in the Pegasus of my imagination, and it was fifteen years later before I dared to mention these kinds of ideas again to my assistants—at the time, Roman Sexl and Helmuth Urbantke, who wrote one of the first serious works about these problems. Since then the number of similar investigations has become legion and reached its apex with "Hawking radiation," which in turn let loose a veritable storm tide of research.

WHAT WAS THE INITIAL SPARK?

A dark energy was the motivating force behind the big bang.

We've seen that, in certain circumstances, negative gravitational energy can become as great as a particle's rest energy. It would then be possible for a particle to pay with negative gravitational energy the ticket price to get out of the underworld and legitimately to conduct *creatio ex nihilo*. We now want to examine more closely in which circumstances this adventure can succeed. If we have an embryonic object with a mass of M, concentrated over a space with a radius of R, then we can bring our test particle with a mass of m up to a distance of R. The test particle will gain a gravitational energy $-GMm/R$, where the universal constant G ("gravitational constant") measures the strength of gravity. As we know, the energy of the test particle is at least its rest energy mc^2. For the leading of the particle to mass M and its creation to remain a zero sum, GMm/R has to be greater than mc^2. This leads to the revelation that it doesn't even matter what the mass m of the test particle is. As soon as $GM/R > c^2$, particles of all masses have a gravitational energy that is more negative than the negative of their rest energy. The universality of gravity means that everything bursts out immediately when the gates to the underworld are opened.

If we multiply the relation above by R, we get $GM > Rc^2$. The inequality can be met by making R of the embryonic mass extremely small. However, quantum theory sets a limit as to how small we can make this. According to Werner Heisenberg, the theory states that a particle, when it's confined to a space the size of R, begins to whirl around with a speed of $v > h/MR$. The universal constant h is called Planck's constant. If the speed of the confined particle would become $v > c$, then confinement would cost a lot of energy. I would have to invest more than the rest energy Mc^2. I could then transform this energy either into kinetic energy of the particle or create more embryonic masses. If I want to avoid reproducing particles, and therefore keep v below c, then the following rela-

Figure 1.4: Max Planck at the height of his fame

tion is necessary: $R > \hbar/Mc$.

The expression $\hbar/Mc$ assigns mass M a length ("Compton wavelength") that is about 10^{-13} m for an electron and 10^{-16} m for a proton. To ensure that a particle with mass m at a distance of the Compton wavelength for a mass M gets a gravitational energy deeper than mc^2, I need to continue $GM > Rc^2$ to $Rc^2 > \hbar c/M$. Both conditions together demand that $M^2 > \hbar c/G$ and define a minimum mass M_P (Planck's mass) as $(\hbar c/G)^{1/2}$; the experiment only works when M is greater than M_P.

For Max Planck, who was the first to introduce M_P, our basic units of measurement of meters, kilograms, and seconds were too arbitrary; they're based on features that the earth just happens to have—its mass, its volume, and the length of its rotation. Visitors from other stars wouldn't be familiar with them, but they would also have to have access to the speed of light c, Planck's constant $\hbar$, and the gravitational constant G. These must even be recognized by a cosmic measurement committee, because it's possible to build a unit of mass, length, and time out of them. We're already familiar with Planck's mass, $M_P = (\hbar c/G)^{1/2}$. Planck's length L_P is M_P's Compton wavelength: $L_P = \hbar/M_P c = (G\hbar/c^3)^{1/2}$. This is both the spatial expansion of the embryo mass M_P, at which it becomes a gravity trap and catches particles from the underworld (its Schwarzschild radius). Finally, there's Planck's time t_P, which is the time light needs to cross

Planck's length $t_P = L_P /c = (G\hbar/c^5)^{1/2}$.

How big are M_P, L_P and t_P really?

Tiny: $M_P = 10^{-9}$ kg, $L_P = 10^{-35}$ m, $t_P = 10^{-43}$ s.

It's no wonder that Max Planck wasn't showered with praise for these figures and that earthlings kept their provincial units. However, times have changed in the last hundred years. Not that there's a store where you'd be understood when asking for a hundred million Planck's mass of flour, but the quantum cosmologists live in spirit in a universe that only has G, $\hbar$, c; they are constantly thinking in Planck units, and that's just what we'd like to do for the rest of this chapter.

Let's return to our alchemic super kitchen in which we want to produce the universe. First we need a mass larger than M_P, concentrated over an area smaller than L_P to tap into the underworld. The question is whether that already creates a big bang. At first glance, the answer is no! The whole thing collapses together in a time period less than t_P. Why does our experiment fail so miserably? It's due to the universality of gravity, which makes everything attract everything else. The equation $GM_P/L_P = c^2$ tells us that we're moving at the edge of a black hole. If the situation becomes so extreme, then there's no way out; everything converges in the middle. You would think that if we gave our embryonic mass a strong enough inner pressure, then maybe it wouldn't be crushed. That this is not the case is shown in Einstein's theory of gravitation. To explain this, I'd like to go back to my conversation with Einstein.

What bothered me then was that Einstein's theory of gravitation began with the following premises: our space-time has a non-Euclidian—that is, warped—geometry, and the metric of this space is the gravitational field. It seemed to me to be logically more satisfactory when the gravitational field could be dealt with like an electromagnetic field, but perhaps a bit more complicated. Gravitation should determine the geometry of space-time due to its universal nature. It should not only produce an attracting force, but also bulge out space to make it appear non-Euclidian.

Einstein wasn't satisfied with my approach. He thought that when you followed this argument, there wasn't any reason to make the gravitational field more complicated than the electromagnetic. Then you should actually take a field with a single component (a "scalar"). Everything else would contradict the commandment of simplicity and be a sin against the Holy Spirit. I didn't see it that way, but nonetheless shamefacedly withdrew my argument. When I later had the time to think through his comments, I realized that, from a purely logical point of view, they were wrong. Even when you argue from the standpoint of geometry as he did, you can characterize a simple non-Euclidian geometry (a "conformally flat geometry") by a single field; such scalar gravitational theories can now be found haunting the literature.

Einstein's instinctual response was nonetheless correct; one consequence of his theory, that the gravitational field actually has a more complex structure, has been corroborated in numerous experiments. This field is described by a quadratic pattern of four lines and four columns (a matrix or tensor of the 2nd degree). Out if its $4 \times 4 = 16$ components, only 10 are independent of another, and each one has its own source. The energy density (also represented by E) is the largest component in the corresponding source matrix. However, the effective source of gravitation in the situations described is $E + 3p$. The letter p stands for the pressure at the specific point in space. In calm situations p is much smaller than E, because p is usually as large as the kinetic energy. In a particle, the energy density E corresponds to its total energy (kinetic energy and rest energy); usually E dominates, and the addition $3p$ doesn't matter. However, already in photons (= light quanta) the pressure p equals $E/3$ and becomes $E \sim p$, when particles move close to the speed of light.

By the way, my feelings of remorse didn't last long, and I published my ideas about Einstein's theory of gravitation. The reception was mixed, as always when an outsider wants to butt in; some thought that this view was wrong, but there was a positive echo from the relevant people like Heisenberg, Dirac, or Oppenheimer. The latter even gave a seminar in Princeton about it called "You Don't Have to Be Einstein to Discover General Relativity." I should emphasize that I did not develop any new theory of gravitation. Every sensible way leads to Einstein; I just wanted to be able to understand a little better what the experts on general relativity were saying.

Returning to our sorcerer's kitchen, we first will see why even making p as big as we want won't stop the gravitational collapse. Greater pressure even makes the situation worse, because the source of the gravitational field is $E + 3p$, and a greater p means a stronger gravitational field. Roger Penrose first derived this statement with all of its implications in 1965 from Einstein's theory. His results announced that gravitation knows no pity when everything is collapsing and when $E + 3p > 0$ is true. Everything will be squashed and crushed together at one point called singularity.

Penrose used the latest mathematical findings of the twentieth century, but unfortunately most physicists were not familiar with them, and he was greeted by a lack of understanding. In his mathematical magic, under certain conditions infinite sequences of mathematic elements will always come close to a border element. This might be reminiscent of the causal proof of God offered by the Scholastics who simply argued that everything that happens has a cause, and this infinite chain of causes must lead to a final cause, which is God. This kind of hocus-pocus makes lots of people feel cheated. After a certain period of time, Penrose's argument became part of common knowledge and is one of the most important statements to come out of Einstein's gravitational theory.

Does this mean that we have to give up for good our alchemic attempts to create the universe? Not quite, because there's a way out of this which was first discovered by Einstein himself.

Not that he was trying to establish *creatio ex nihilo*; that's not the kind of thing he would have ever done. He was exploring the seemingly hard fact of the unshakable permanence of the heavens. This point of view has been proven wrong in the meantime. The apparently eternal constancy of the cosmos conflicts with gravity, which is always looking to make things collapse. We apparently need some "antigravity" to counteract the effects of gravity. To spice the text up, I've used a word from the physics popular press, where all sorts of crackpots believe they know everything. How can antigravity even exist when we follow the rules of the game in Einstein's theory?

Naturally, Einstein didn't have access to Penrose's article, but we know that antigravity requires a situation where Penrose's energy condition $E + 3p > 0$ does not hold.

How can we achieve this?

For matter, energy E (kinetic and rest) is always positive. Negative energy remains the privilege of gravity.

How is it though with negative pressure? This was once called an "inner pull" by Schrödinger and might seem at first to be nonsense.

Einstein, however, was clever and knew that only changes in pressure can be measured. At first glace, a constant negative pressure doesn't contradict experience, although it goes against our intuition. The discerning reader will once again be tempted to cry foul, but stay with me. We'll be coming to better explanations later.

Einstein introduced into his equations the so-called cosmological constant and meant an energy evenly distributed throughout space and an equally large but negative pressure. The word *cosmological* just means, "Don't worry. You won't notice this in your everyday life anyway, I just need this fudge factor for my cosmic tricks."

For $p = -E < 0$ is then actually

$$E + 3p = -2E < 0$$

and there we have our antigravity with which Einstein wanted to stabilize his universe. However, it wasn't really all that stable as the balance between gravity and antigravity is too delicate. A little too much gravitation and the universe collapses; a little too much antigravity and it breaks up. Then there's the experimental evidence that Einstein began with the faulty assumption of a stable universe: our universe is exploding; it's not static!

The pinnacle of this unhappy situation was in 1920 when the Russian physicist Alexander Friedmann found solutions to Einstein's equations that allowed for the expansion of the cosmos without needing the cosmological constant. To explain that the universe is still expanding, we can simply assume that everything flew apart in such a way at the zero hour that gravity could no longer stop it. Enraged, Einstein threw out his cosmological constant and called it the biggest blunder that he'd made in his life. Maybe he did pine for it later? Throughout the year that I spent in Princeton, I saw Einstein only once in a physics colloquium. There was a discussion about cosmic expansion, and Einstein got excited and wanted to know if it was decelerating or accelerating. The data at that time couldn't give him an answer. If only we could see the look on Einstein's face if he could know that it seems to be the second alternative! The cosmological constant is the leading force in cosmic expansion, and Einstein would have to say that his biggest blunder was ever having doubted it.

Perhaps the current acceleration in expansion is caused by the fact that there is something left of the cosmological constant after the inflationary phase, which we will be discussing shortly. There, acceleration is evidence of a background of positive energy and negative pressure called "dark energy," which is one of the least understood phenomena in modern physics.

According to current quantum field theory, even the most perfect vacuum has an energy density ("zero point energy"). Each particle corresponds to a field and, according to quantum field theory, this field has a minimum energy, a particle's rest energy per (Compton wavelength)^3. It also provides an equally great pressure, but with the opposite sign. This would be a gravitational source like the cosmological constant, but it's much too strong. Whichever particle you take, a rest mass per (Compton wavelength)^3 produces an enormous energy density. The Compton wavelength for a proton, for example, is about 10^{-16} m. A proton mass in a cube of this length would produce a significantly greater energy density than that of the universe; there you find on average only a few protons per m^3.

It's possible to come up with this cosmic density of matter in the following way: a typical star contains 10^{57} protons and the average distance between two stars is about 1,000 light years, around 10^{19} m. If the protons in a star are distributed in a cube with sides this long, there remains roughly one proton per m^3; $(10^{19})^3 = 10^{57}$.

We have to be very exacting in getting rid of the zero point energy generated by the quantum fields, whose density in order of magnitude $(10^{16})^3$ = 10^{48} times greater than the actual density. We've got to be careful, though, in cutting it out because we need about five times as much as the visible energy for this "dark energy." How exactly this should happen can't be explained by even the cleverest thinkers.

Although disowned by Einstein, the cosmological constant has survived into the twenty-first century and is today subject to enthusiastic debate.

The first good usage came from the Dutch astronomer William de Sitter. He constructed a model of the universe that can be called pure antigravity. It first collapses on itself but then is taken apart by antigravity and is scattered to eternity. The de Sitter universe only has antigravitational matter $p = -E$, where both quantities remain constant, which offers several advantages. As this kind of universe will prove to be quite important for our purposes, we should explore some of its features.

1. Greatest Possible Symmetry

A space is said to have symmetry when it appears the same when looked at from different angles. The world we live in is isotropic, meaning that all directions are equal. The fact that we think that above/below, east/west, and north/south are different directions is based on the earth's influence; far out in the universe these distinctions don't exist. Our space is also homogenous, meaning that all points are equal. All other opinions are pure provincialism. It was Einstein's great achievement to identify further, more extensive symmetries. They connect phenomena that we encounter as being completely different from each other, like space and time.

When we observe events from a uniformly moving frame of reference, we can mathematically describe them with an intermixture of space-time coordinates. Einstein's special theory of relativity tells us how they need to be mixed so that the laws of nature hold true in exactly the same way for this system. Due to this symmetry, space and time are welded together to a four-dimensional space-time. The original symmetries mentioned can be summarized by the statement that space-time is four-dimensional, homogenous, and isotropic. The de Sitter universe boasts an equally great level of symmetry as space and time in the special theory of relativity, but it took several years for this to be widely recognized. It was here that Einstein made his only conceptual mistake; this had to happen to him of all people. The thing was that Einstein reacted negatively to de Sitter's work; he had the impression that the de Sitter universe could not be our universe. His instinct was correct; we don't live in a de Sitter universe. He was looking for reasons why not and came up with a very flimsy argument. De Sitter wasn't able to describe his universe using a single system of coordinates similar to the way that we're unable to include the entire globe on a single map. With de Sitter, singularities appeared at the edges of the area covered, which Einstein took to indicate a distribution of singular masses, the completely wrong interpretation. If someone had recognized that the de Sitter universe was homogenous, then it would have been immediately clear that the idea of a singularity had to be nonsense: every point is equally good, and singularities can only come from the cosmic maps used.

De Sitter's universe is even temporally homogenous. The inversion point of time, when contraction turns into expansion, appears at first to be distinguished, but isn't. Different frames of reference have different temporal inversions. The de Sitter universe came back into favor in the previously mentioned "steady state" theory, and people wanted a universe that appeared the same from eternity to eternity. It was kind of an act of defiance against the Bible. However, the evidence that the universe had a beginning is simply overwhelming, and this theory had to be dropped. As we will now see, the de Sitter universe was able to be a part of cosmic evolution after all.

2. *Greatest Possible Expansion*

According to the following, an expansion must be exponential when its rate remains the same at every point in time. If $R(t)$ is the radius of the universe at a random point (t) in time, and R increases tenfold after every time unit, then $R(t + 1) = 10R(t)$ for every t. This equation holds when

$$R(t) = 10^t R(0)$$

as then

$$R(t + 1) = 10^{t+1}R(0) = 10 \times 10^t R(0) = 10R(t)$$

for all times t. The same growth rate for all times means an exponential law for time dependence.

The circumstances under which such a law holds true in Einstein's theory were first investigated by Friedmann for a spatially homogenous and isotropic cosmos. In this theory, the energy E is the source of gravity, and this pulls everything together, including the universe as a whole. Pressure p also adds its two bits, and if it's sufficiently negative, it can transform contraction into expansion. Appendix 5 sketches out how in Einstein's theory E and p together cause an expansion as soon as $E + 3p$ becomes negative. Then we can see why we can really use the de Sitter universe for our *creatio ex nihilo*. If an antigravitational situation occurs in the gravitational area that surrounds M_P, then it would expand from L_P to an imposing cosmos in no time. If in the beginning the germ of the world had only $R(0) = L_P = 10^{-35}$ m, then after $t = 35\ T_P$ the radius of the cosmos would already be

$$R(t) = 10^t R(0) = 10^{35} \times 10^{-35}\,\text{m} = 1\,\text{m}.$$

To reach the size of the present universe we would only need 18 more Planck time units, $R(35 + 18) = 10^{18}$ m, which is just 15 billion light years in meters. Our time unit $T_P \sim$ roughly equals 10^{-43} s, and it would only take $t = 35 + 18 = 53$ Planck time units. Even when we round up 53 to 10^2, it's still just 10^{-41} s, an unfathomably short period of time. This might appear to the reader to be progressing at a dizzying speed, and the reader would be right to object that this is not only happening "in no time," but even faster than the speed of light > c. Our c equals one in the Planck units, and as $R(1) - R(0) = 10 - 1 = 9 > 1$, the speed can't always be less than one. For something the size of the radius of the universe, there's no prohibition on going faster than the speed of light. This normal prohibition applies only to localized phenomena as encountered in everyday life. It's also the case in Einstein's theory that *Quod licet Iovi non licet bovi* (meaning "What is permitted to Jupiter isn't permitted to oxen," or people aren't allowed to do what God can). This in no way implies that back then some random matter contested light in a speed race. The superior force of gravity presented everything in existence with unimaginable wings and dictated the permissible speed limits.

The speed of the expansion of the universe can be understood as follows. The substance with negative pressure (dark energy) fills all space evenly. The larger the space becomes, the more of this stuff there is and the more powerful the repulsion becomes. The resulting speed limit infringement solves one of the paradoxes of cosmology: when we look in the distance to the right, say 10^{10} light years, and then equally far to the left, then we're seeing parts of the universe that could never have communicated with each other. One ray of light would need 2×10^{10} years to go from one part to the other, and the universe isn't even that old yet. We still see the exact same conditions on both sides—for example, the exact same temperature value of background radiation. How were the two parts able to agree to be so similar? Einstein's theory offers a simple answer; they were hatched from the same egg but were then separated from each other at a speed that appears faster than the speed of light to us today.

We have nudged the *creatio ex nihilo* into the realm of possibility, and now we just need to find out two things:

A. Where did the "inner pull" (negative pressure, $p < 0$) come from?

B. How can we turn it on and off in order to be able to get on this speeding train and jump off again?

With regard to A, according to Einstein, the sources of our gravitational hole have ten different sources. If there was such great symmetry in the beginning like in the de Sitter cosmos, and it was also respected by the sources, then according to the special theory of relativity, $p = -E$ must be the case. Then an antigravity situation emerged. This is some more of Einstein's hocus-pocus.

With regard to B, if this blaze spread so quickly, then normal material $E + 3p > 0$ was also created, that then overpowered antigravity and reined it in. If we want to get rid of antigravity and lead events to a more regular path, then we would have to break the symmetry of the de Sitter cosmos through normal matter.

The architects of this "inflationary universe" (K. Sato, A. Guth, A. P. Linde, among others) have come up with several mechanisms that allow us to get out of this situation. I can't get into this cosmic technology here; I just wanted to work out some of the ideas of the inflationary universe. It is the preferred image of the creation of the universe today, particularly as it clears up other paradoxes in cosmology. It shows how it's possible to understand the beginning of Genesis scientifically.

GLORIFICATION OF GENESIS

The words of Genesis parallel our modern understanding of how the universe was created.

To summarize our physical understanding of today: matter has the condition of nonexistence. In physics this is called a "vacuum" and is characterized by energy equaling zero. For gravitation, which dictates the geometry of space-time, the analogy would be the nonexistence of space-time. However, gravitation's energy is negative. This means that in quantum gravitation, the vacuum, when brought together with matter, becomes unstable in the face of the building of "small big bangs": it is energetically possible for a Planck mass of matter, concentrated on a Planck length, to come into being by itself. Of course, this is repressed and delayed by a Gamow factor, but as long as there's still no time yet, it doesn't matter how long there is to wait. The vacuum is crackling with these sparks, and if one is challenged by antigravity, it proliferates rampantly and becomes a big bang. Then it's necessary to exit this de Sitter phase as soon as possible to reach the kind of universe we want. If the expansion lasts too long, the cosmos would become a bleak wasteland. The real miracle in the first day of Creation is in the comment "and God saw that it was good." Up to now we have seen that to create some kind of universe is easy, if not even necessary. To get a universe that suits our purposes requires amazing precision. The characteristic time was $t_p = 10^{-43}$ s, and our cosmos is 10^{10} years old $= 10^{17.5}$ s, or 10^{60} Planck time units old. Its natural life span was, however, expressed in Planck time units and would be in its range. The world is therefore outrageously old. Not many of the ten billion years given as the age of the universe can be bartered away if we are finally to come into being. Biological evolution requires 10^9 years. If the big bang had been too weak and everything had first collapsed upon itself again, then we wouldn't exist. If it was too violent, then everything would thin out too fast, and there wouldn't be elements vital for our survival like carbon or oxygen. They are not created at the big bang, but are rather slowly bred in stars and make their public debut only when the stars explode. If the cosmic gases that stars can come out of became too diluted after the first generation of stars, then there would be no descendants. We, however, are the offspring of a later generation of stars. The development of aristocratic beings like us takes time.

The creation of the universe is like the launching of a satellite around the earth. If you don't use enough propellant, it will fall right back down; if you use too much, it will fly away into space. It took several attempts in aeronautics to get it right and to find a stable orbit around the earth. There was an analogous situation at the big bang; just the demands on the exactitude were incomparably greater. They corresponded to a precision of 10^{-52} m for an orbit around the earth, beyond human comprehension. This should have happened just by accident? How absurd. Even if you wanted to aim exactly for a ten-billion-year-old universe from the very beginning, you couldn't. Then came the theory we just discussed about the inflation of the universe, and in the blink of an eye, the cosmos was finished in $53 \, T_p$ and all those exponents overcome. In this way it's possible to explain the course of the universe, but not to predict it. The reason for this is that it's possible to get off this dizzying ride only through a tunnel effect like with the a decay. In radioactive decay, the probability of burrowing through the electric fence is equally great at every time. No one can prophesize when the burrowing through and, thus, when the decay will take place. To bring about a useful cosmos, we have to catch exactly the right time. We are once again dependent on coincidence, which has already been very well disposed toward us.

Let's now compare the scientific picture of events with Genesis.

In the beginning God created the heaven and the Earth.

Today we would perhaps identify heaven and earth with gravitation and matter.

> *The Earth was without form and void, and darkness was upon the face of the deep; and the Spirit of God was moving over the face of the waters.*

The vacuum state, unstructured and empty, dominated matter as well as light. There weren't any other spirits beyond God.

And God said, "Let there be light," and there was light.

The language of God is the laws of nature, and according to them, the vacuum in the quantum gravitation is unstable. This instability can develop into a big bang, an avalanche of light that creates our immense universe out of nothing. Light actually does dominate in the beginning in our modern models of the universe, and it is only much later that matter develops into the leading energy carrier.

I want to emphasize that I'm not presenting all of this with prophetic zeal, but rather just as a possible line of thought offered by modern science. The facts can't prove the existence of God, explain him, or make him comprehensible. Such a choice of words would only point to an overestimation of the human ability for understanding. Today we have a much more grandiose and cerebral image of the Creation, such that the story of the making of the earth six thousand years ago appears somewhat pathetic in comparison. I like best the following, poetical choice of words found in Isaiah to describe the knowledge given to us:

We have seen the hem of his robe.

2

Is Everything Just Random Coincidence?

THUS SPOKE NIETZSCHE

"God is dead," cried Friedrich Nietzsche in desperation.
"Are we not straying as through an infinite nothing?
Do we not feel the breath of empty space?"

THIS FAMOUS REMARK took on a life of its own, and an enlightened and self-confident society took it up almost as a battle cry to proclaim the triumph of reason over faith in revealing the secrets of nature. This society was intoxicated from the pioneering advances in science made over the centuries, be it in biology, physics, or chemistry; and the conviction that the regularity of phenomena in nature could be explained by a few axiomatic principles became something of a dogma. Confrontation with church teachings, and indeed with the belief in God himself, was actively sought. When physical-chemical or biological events could be understood as following a set of laws, then "naturally" the Creation could be explained, and there was no longer any need for God-the-Creator.

I will now attempt to explore the idea of "naturally explainable" and to determine in what sense the laws of nature explain the evolution of the universe.

First, naturally explainable should be that which doesn't contradict the rules of science. Does this mean that the existence of a living God would need to manifest itself in having God break one of the laws of nature he created in a display of power? Because according to this point of view, there's no need for a God who obeys the rules he made. To check this thesis, whose logic appears to me to be contestable, we first need to be clear about what the known laws of nature can do and when too much is expected of them.

What Do the Laws of Nature Control?

The known laws of nature are a combination of randomness and necessity.

The basic laws that determine the evolution of a physical system relate to two aspects: on the one hand to the "state" of the system, consisting of all of its features initially accessible to us, and on the other to its "dynamics," the changes in the state over time. The "system" is used here as a general term and sometimes refers to the whole universe. The state of a point particle is determined simply by stating its position and speed; for when many particles are involved, the situation becomes accordingly more complicated.

The dynamics reflect the unwavering character of the laws of nature. They are rigidly deterministic: if we know the state at the beginning, then we can precisely predict how it will look after a specific period of time. Generations of physicists have worked on the equation for motion until they were sure that, for the initial state given, there was one and only one solution. This means that, for any future point in time, exactly one state was predicted. The initial state itself introduces an element of randomness to events. It has in no way been laid down by the laws of nature. It's like in a card game. Whichever card we get is determined by chance in the shuffling before the game; the rules of the game then say what can happen next. How the rest of the game will develop also depends on the decisions taken by the players. Similarly, the laws of motion in quantum mechanics simply regulate the different possibilities; we determine by our choice of measurements which state will be realized. The mathematical description of this kind of dynamics can best be imagined as the states spreading out in a (high-dimensional) space, the "state space." Every point in this space corresponds to a state defined as exactly as possible. You are free to choose which point to begin with, because this is not dictated by the laws of nature. The dynamics lay out a curve through this point. It shows how the point moves in the state space over time and is called the orbit, or, to be more scholarly, the trajectory. As the initial state hasn't been previously determined, it must first be measured. Measurements can only be finitely exact, so that we can only approximately know the initial point in the state space. We just know that it was in a certain area that we then call "state." In the comparison with the card game, we're in the position of the audience who can't see all the cards. Therefore, the predictive power of the equation of motion in chaos systems vanishes over time. In these systems the orbits originating in one area go their own way; they travel apart exponentially. If the distance doubles in time *t*, in 2*t* it will have quadrupled, in 3*t* octupled and so on. For weather forecasts *t* is about a week; for the orbit of our planetary system *t*, called the Liapunov time, is more than ten million years. There will come a time when the web of orbits originating in one area spreads over the entire state space like a foam. Afterwards nothing more can be known about the original state. This is demonstrated in Figure 2.1 in simplified dynamics:

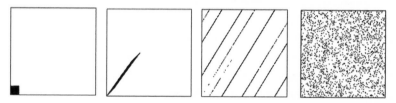

Figure 2.1: The development of a small area in the state space according to a chaotic law of motion. The left-hand picture represents the initial state (t=0). The following pictures are the result of one, four, and eight iterations of the illustration.

Here the state space is a periodic square; the dynamics expand in one direction and compress in the perpendicular. Periodic means that, if you come out on one side of the square, the periodicity leads back into the opposite side. In Figure 2.1, we first know with a high degree of certainty that the initial state was in the lower left-hand corner, but this certain knowledge fades over time until our information is apparently distributed randomly throughout the square. In reality, there's a complicated plan behind the distribution. It says that the points will gather together again when the dynamics run backwards. However, this evidence of order remains hidden from our view.

I would like to describe the situation in chaos systems using the following language: a state can be explained (or not, as the case may be) when there is an initial state according to which it occurs. It is predictable when it will occur relating to most of the initial circumstances. Something can therefore be explained when it can occur keeping with the dynamics (the "laws of nature") and is possible according to the laws we know; predictable means that it has to be this way. The laws of nature make it so.

With this terminology it's possible to characterize chaos systems as follows:

After a certain period of time, you might be able to explain everything, but nothing can be predicted.

This certain period of time is the Liapunov time just mentioned. How long it takes until the predictive powers of the laws of nature disappear depends on the physical features of the system and the dynamics. For every chaotic dynamics there comes a time after which it's impossible to say what was predicted by the initial state and what exactly the dynamics were. For all chaotic dynamics the initial area will ultimately expand over the entire state space. Therefore, the same thing is to be expected for every other measurement, which will also be only finitely exact and could lead to various chaotic evolutions.

These insights shake the argument set out in the beginning as we can't determine if a chaotic evolution over millions and millions of years has broken a basic law. It could have been following other rules to come up with the same result to which we have access. To show this, we've replaced the dynamics in Figure 2.1 with a somewhat more expansive illustration. It's possible to see after the first lines in Figure 2.2 (opposite) that, after eight iterations, basically the same picture is created as before. To discover if another law governs here, in the last picture the inverse dynamics of Figure 2.1 are used up to eight times. The configurations in the second line of Figure 2.2 correspond, from right to left, the times eight, four, one, and zero of the initial dynamics. We see that this does not yet gather the points and make sense out of the chaos. You can only make sense of something when you know the exact law for development.

Therefore the reasoning that God is dead because he sticks to all the rules not only uses questionable logic, but is also not verifiable. However, the macho-bully God who breaks his own rules whenever he wants is also unnecessary. We will see that the invisible God can act much more miraculously than through acts of force, even when some things remain beyond

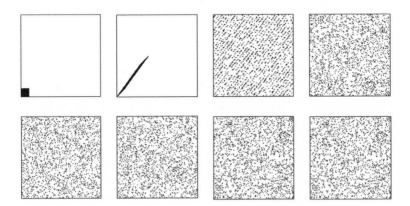

Figure 2.2: First line: temporal development of a chaotic illustration with a rotation about twice as strong as in Figure 2.1. The pictures from left to right once again represent zero, one, four, and eight iterations. Second line: from right to left the inverse of 2.1 progressively four, seven, and eight times is used on the final state reached (at time t=8 on the right-hand side), representing the times four, one, and zero.

human comprehension. We would also not interpret a violation of the laws of nature by some sort of display of power as such, but would rather say that we've got to revise the laws in this instance. That would mean creating a random patchwork. We, however, are witness to a grand scheme of order in our universe.

Randomness and Necessity

Can observations on the probability of randomness during Creation give us new insights?

In his worthwhile, if somewhat dogmatic, book *Chance and Necessity* (*Le hasard et la nécessité*), Jacques Monod made clear that both played a role in the origins of life. Some things were formed by necessity, but many developments, while not breaking the laws of nature, could have progressed differently. In our terminology, these developments can be explained, but not predicted; they are random. There are many who would be shocked and outraged to hear what they consider blasphemy, along the lines of "you're telling me that human beings, the pride of Creation, are only random products of coincidence?" In the meantime, it seems to us that biological evolution is just the tip of the iceberg. The development of the entire universe is determined by these random events; we human beings are in good company. We will explore the individual stages of the cosmic evolution in later chapters and investigate how accidental the individual "frozen-in accidents" of Murray Gell-Mann really are. Taken separately they are not inexplicable, but taken together they really are quite remarkable.

What exactly do we mean by randomness? It does not break the laws of nature when, in roulette, someone bets on seven and the ball rolls to seven. If the player bets on seven again and the ball rolls to seven again, we say that he's very lucky. If this happens ten more times, then we would say that can't be a coincidence. If the player had a flair for the esoteric, then we'd say that he'd performed a miracle; if not then we'd say he must have cheated. After having seen that random events occurring during the development of the universe were always in accordance with whatever was needed for humanity to come into being, we could ask ourselves if God didn't perform a miracle and set the signs at all the crossroads in such a way so that his image could appear. Of course, a very limited meaning of the word "image" is to be understood, but before we get carried away by such things, we should examine what is found behind the term "random event." A random event is something that has happened without the laws of nature (known to us) demanding it to. This term therefore also seems to make a statement about the current stance of science.

The statement that evolution has happened to take a fortuitous turn doesn't say much, because it had to make a turn, so why not a good one? It would make more sense to say that evolution has taken an improbable turn. We're not really interested in a particular turn of events. There are certainly many equally good, and we would be satisfied with any of them. For example, it doesn't really matter to us where the earth is on its orbit, which is almost circular, around the sun right now. The important thing is that we are neither too close nor too far away, so that it's not too hot or too cold. The probability of such a class of good events equals: (number of good events) / (total number of events). This is a useful term for us, particularly as the total number of events in our physical systems is immense, but in a finite universe also finite. Its finite nature is a result of quantum mechanics, in which the state space has something like a grainy structure. In the universe all possibilities provided for by the laws of nature should be exhausted, because in the beginning it was unimaginably hot, and statistical physics tells us that then all possibilities are equally probable.

A GAME OF MARBLES

We will now direct our musings on probability toward a simple concrete example. Let's say that we had N numbered marbles and each one just had one of two characteristics, let's say "white" or "black." Then for two marbles there are four possible combinations: (white, white), (white, black), (black, white), and (black, black); for N = 3 there are eight possibilities, in general 2^N possibilities for the group of marbles. Let's assume that the best possibilities are those with a certain order. Order is the opposite of disorder—that is, having everything mixed up or 50 percent black, 50 percent white. For N = 2, order and disorder balanced each other out: there were two monochrome and two mixed possibilities. For N = 3 there are still two monochromatic and thus 6 = 8 − 2 mixed possibilities. For a larger N there is a sharp increase in the number of possible color combinations having half of the marbles black and the other half white.

To put it simply, there is much more disorder than order. These considerations will now become somewhat mathematical, and I don't want to bother all the readers with them. I'd like to discuss them further in appendix 6. Right now I'm just going to present the results and ask for your patience. In general, the number z_n of possibilities for combinations of n marbles of one color and $N - n$ for the other is given in the following equation:

$$z_n = \frac{1 \cdot 2 \cdot 3 \cdot \ldots \cdot N}{[1 \cdot 2 \cdot \ldots \cdot n][1 \cdot 2 \cdot \ldots \cdot (N - n)]}.$$

The probability w_n for the colors $(n, N - n)$ is according to our definition the probability

$$w_n = z_n / 2^N,$$

as 2^N is the number of all possible color combinations. The w is the largest for the mix $n = N/2$ (greatest disorder) and sharply decreases with increasing order. If d is the deviation from 50:50, meaning $n = (N/2)(1 + d)$, then, when we describe w instead of n with d, we get

$$w_d = 2^{-Ncd^2} w_0,$$

where the number c is approximately equal to one. (The calculations leading to these equations can be found in appendix 6.) In any case w_d quickly decreases as N increases. Even when cd^2 is only 1/1000, meaning that n is extremely close to $N/2$, then for $N = 10^6$ the probability w_d is already approaching zero. w_d is only w_0, the probability of $n = N/2$, divided by almost 2^{1000}. This number would fill many lines if it were written out. It was the act of genius when Ludwig Boltzmann introduced the logarithm of the number z_n of possibilities as a measure of disorder and recognized that this unit plays a central role in thermodynamics under the name of "entropy." It is usually represented with the letter S, and its exact definition can be found carved onto Botzmann's gravestone:

$$z_n = e^S.$$

We'll meet entropy again in chapter 6, but in a simplified form.

How Improbable Is Order?

One of the secrets of the universe is the spontaneous emergence of orderly structures.

Our observations have shown that with a large number of N independent parts, the greatest probability is that disorder will prevail: the expected result. It then follows that the probability is extremely small at first that the orderly conditions that prevail today in our universe could have developed out of its initial chaos. According to quantum mechanics, we should replace the N of our simple example with the number of atoms; 10^{24} for a body our size and 10^{80} for the entire universe. It therefore wouldn't help to point out that there are 10^{11} Milky Ways each containing around 10^{11} stars for a total of 10^{22} stars, and on one of them surely order would happen to develop. Despite our bounty of stars, this wouldn't happen; w_d is simply too small. Even when $d = 1/1000$ and $N = 10^{80}$ then

$$w_d = 2^{-10^{75}} \sim 10^{-10^{74}}.$$

The number of stars needed would have 10^{74} zeros; the number of our stars has only 22, and 22 is nothing compared to 10^{74}. We can see that our a priori probabilities are hopelessly off the mark. We've got to find some other angles.

One of the crucial features in the dynamics of chaos systems is ergodicity, also a subject of Boltzmann's research. This says that systems run through all of the possibilities available over the course of time, and, in the time average, all of them the same number of times. Thus, the probability of a good possibility as discussed above should be able to tell us how often it would occur in the time average. In large systems we would again then only have a minuscule sliver of the time when order is to be found. However, as there are so many possibilities in large systems, the time needed until all of them have been covered is much greater than the age of the universe. Therefore, such a time average is irrelevant for us. This raises the question of whether there isn't a shortcut that would allow only the better possibilities to occur first and bypass the less favorable. By the time the less favorable possibilities would emerge, they wouldn't matter to us. At the level of generality at which we've been arguing, it's not possible to decide if this is the case. It is, however, conceivable; in biology, for example, to say that natural selection sticks to a path that was once chosen, without bothering with ergodicity.

In the following chapters we will discuss random events in individual stages of the evolution of the cosmos. It is up to the reader to decide how random these seem. There are always clever people who conceive of ingenious mechanisms to explain them. Most of these theories don't hold up and are thrown out. When a model does work, it's published in triumph and felt to be completely natural. In chapter 6, pages 134ff, we'll find laws that must produce order. These are, however, just caricatures of the laws we take to be fundamental. It's hard to say how well they represent the latter; perhaps it's just a "coincidence" that they "happen" to work. It would be foolish to attribute everything to chance. Great discoveries in science have been made when someone further explored what generations had taken to be a random occurrence. The root of Einstein's important theory of gravitation is the equality of inert and heavy mass, which had up to that point been accepted as a mere brute fact. Einstein had already recognized that there was something deeper behind this and then revealed to us that the actual nature of gravity was the geometry of space-time. This geometry determines how effects can propagate and gravity came to rule over causality. The moral is not to approach the wondrous facts in the cosmos with your mind already made up and simply reject other interpretations by giving them currently unpopular labels.

At the end of this chapter I would like to relate three episodes of my life showing how I encountered all sorts of ranges of sizes. They illustrate not only randomness and necessity, but also the caprice that scientific evolution can have.

LIMITING THE ORDERS OF MAGNITUDE INCREASES THE OVERALL VIEW

The foundations of modern science were not understood
for a long time.

The exponential calculations sketched out in chapter 1 regarding the size and age of the world had fascinated me already when I was a schoolboy and led me to my first scientific experiences. Not that I can recall in detail what we learned in science in our school, but I still know the school's address by heart: Vienna 19, Alfred Wegenergasse 10-12. I had often asked myself who this Alfred Wegener was, who at least had a side street named after him, and I finally got the answer that, "He was a professor in Graz, a polar explorer and had a fatal accident in Greenland. He also had the crazy theory that the continents could float across the earth's surface like ice floes, but that's just nonsense." I was nonetheless fascinated by this idea, and it seemed to me that, if the earth's core was liquefied, then it could really have something like ice floes. I thought of a picture of an earthquake where the earth opened up wide enough to put your hand through, and I did the following calculation: if this kind of earthquake happens somewhere once a year, then the earth's crust shifts 10^{-1} m a year, or in a hundred million years $10^8 \times 10^{-1}$ m = 10^7 m = 10^4 km, or ten thousand kilometers. That is easily the distance between Europe and America, so that in this time continents could very well have grown together or separated and spread apart; it seemed random how. When I tried to talk about my thoughts with others, I was told that I would do better to leave such numerical games alone and trust the experts. It's a fact that Wegener's continental drift theory was not only rejected at our school then (in 1938), but even in my 1967 edition of the *Encyclopedia Britannica*, there's no mention of him. I could never understand the psychological reason for this rejection. Perhaps the fairy tale of a six-thousand-year-old universe was still swirling around; according to this you'd only be able to move 600 m and not make it to America. Today, continental drift is the dogma of a complete branch of science—plate tectonics—it's seen as the division between superstition and profound knowledge.

IT CAN LEAD TO MISUNDERSTANDINGS IF ONE PERSON JUST WANTS TO KNOW THE RANGE AND THE OTHER AN EXACT NUMBER

Scientific results can be viewed from different angles.

At an anniversary ceremony, I heard two lectures praising the high level of physics in Austria between the wars. The theoretical physicist emphasized that the best quantum theoretical calculations in nuclear physics were made here, and the experimental physicist proudly pointed out that the most exact measurements of radioactivity were made in Vienna. During the break, I asked my friend Willibald Jentschke (the founder of the German electron synchrotron DESY), "Tell me, Willy, if you were all so good, then why were the major discoveries made elsewhere?" He replied that there were difficulties in the contact between the theoretical and experimental physicists, and he relayed the following mini-drama:

A theoretical and an experimental physicist were working on the same problem independently of one another and happened to meet.

Theoretical physicist: Esteemed colleague, I was just successful in calculating the life span for the a decay for substance X, and it's a month.

Experimental physicist: Sorry, but that's completely wrong. I have measured it myself exactly, and it's one month, two weeks, three days, and five hours.

Theoretical physicist: I don't care about that exact figure anyway!

Whereupon the experimental physicist leaves the scene, outraged by the theoretical physicist's attitude, and they never speak to each other again.

To fully appreciate this story, you have to imagine what a decay looks

like. The substance from which heavy nuclei are formed has a lumpy structure; the smallest lumps are identical to the nucleus of the element of
helium. Due to historical reasons, the accepted term for such a lump is a
particle. As they also have a positive charge, they are repelled by the
nucleus and ultimately flung away. The a particle whirls around in the
nucleus at an incomprehensible speed and beats 10^{22} times a second against
the nucleus' surface (from the inside). Perhaps only after 10^{28} attempts, it
can happen to find a leak and escape. Sometimes it takes much longer. Of
course, it's not possible to calculate precisely gigantic numbers like these
and the Gamow factor using simple models. To be able to predict the leap
of 10^{-22} s between two escape attempts until success after 10^6 s is a great
achievement in itself, which the experimental physicist did not catch. He
probably wasn't familiar with Einstein's wise words, "better to be approximately right than precisely wrong" and was focused on the many uses of
radioactivity, the foundation of which is knowing precisely how long the
nucleus' life span is. The theoretical physicist didn't care about these practical considerations, and so they couldn't relate to each other. The big loser
was the scientific climate.

If You Want to Know Too Much about Something, You Can Stir Up Muddy Waters and Reap Little Thanks

Sometimes small improvements in calculations can have a huge impact.

To conclude our experiences with large numbers I would like to relate my
first meeting with Wolfgang Pauli. I had the good fortune to work as his
scientific assistant at the ETH in Zurich from the fall of 1951 until the
summer of 1952. At twenty-four, I had already published a few small scientific articles, but compared to the great Pauli I was a complete nobody.
He never made me feel that way, but always treated me as his equal. This
doesn't mean that he was saccharine sweet, as he, for example, once said
to a colleague, "Mr. X, I don't mind that it takes you so long to think; I
just mind that you publish faster than you think."

**Figure 2.3: The protagonists Angas Hurst, Wolfgang Pauli,
and Walter Thirring in 1952**

Pauli left it up to me which problems I wanted to solve, and I wasn't exactly modest in my choice. At that time Pauli and Heisenberg had paved the way for a development that had become a certainty, and it would transform our understanding of the nature of matter. According to it, matter is subject to a constant emergence and disappearance. Beyond our visible world, there's a kind of world of shadows (called the underworld). It fills all space and contains the plan for all atomic particles. These don't exist there in actual form, but only as possibilities. When enough energy is invested, particles can be transported out of the underworld into our world, just as the a particles break out of the atomic nucleus. Even if you're poor and can't afford a high-energy machine, an occasional visitor from the underworld will stop by (virtual particles). For the electron these are photons, positrons, and other electrons, which swirl around it about 1 percent of the time and shake its magnetic moment. This is then changed from 1 (in suitable units) to 1.001 (first generation). These visitors are accompanied by others, who change the value by 10^{-4} (second generation). At that time there were precise measurements taken that confirmed the effects of the first generation of visitors, and ever since the experimental and theoretical physicists have been spurring each other on. The experimental physicists make more and more precise measurements and force the theoreticians to calculate the influence of further and further generations. The theoreticians do it and thereby force the experimenters to make even more precise measurements until today we have the magnetic moment, μ of an electron as

$$\mu_{Exp} = 1.00115965219 \pm 0.000000000041$$

$$\mu_{Theor} = 1.00115965219 \tag{2.1}$$

and have achieved the most exact agreement between theory and experiment in physics. The ± quantity represents the possible error margin in the measurements. The discerning reader will find it a bit suspicious that there's such a large degree of agreement between theory and experiment and might think that there's something fishy going on. To certain degree that's true. μ_{Theor} is dependent on natural constants that cannot be exactly measured individually. The experimenter takes μ_{Exp} as a measurement of these constants and adjusts them so that μ_{Exp} and μ_{Theor} form a perfect match. Some of my colleagues find this approach to be a bit amateurish and prefer to use the average of other measurements for these constants. This changes the last decimal place slightly, but the level of agreement remains impressive. It shows how precisely the picture of the visitors from the underworld predicted by the theory was right, despite quantum mechanical fuzziness.

I wasn't so fascinated by the exact figure at that time; it seemed to me to be more of an amusing pastime for simpler minds. For me, it was the principle of the thing, the image of the visitors from the underworld. Would the effects of following generations become smaller and smaller, or could they snowball and cause an avalanche? I had the feeling that something like this would be possible and wanted to come up with the mathematical proof. The time and energy needed to calculate the magnetic moment of an electron was much too great, so I chose a simpler characteristic of a hypothetical particle that had the same mechanism. The further chronology of events is repeated in telegram style not to waste too much space on a side topic.

December 1, 1951: I'm agonizing over how to get to the root of further generations when it already takes months for the best to calculate the second generation.

February 1, 1952: Trash can filled with mountains of calculations, but no real progress.

February 10, 1952: The contributions of even the tenth generation are anyway so tiny that I'm already far beyond what can ever be measured.

February 15, 1952: The effects are only still 10^{-50} in the fiftieth generation.

February 20, 1952: I'm able to prove that, even if I can't precisely calculate the effects of the nth generation, this number must at least be larger as a positive number specified by me.

February 21, 1952: Interesting! At around the sixtieth generation the diminishment of their effects begins to slow down.

February 22, 1952: Pauli enters the room, "Mr. Thirring, the ETH is going skiing next weekend; wouldn't you like to come?" I answer, "That would nice, Professor Pauli, but I would rather finish this work here." Pauli says, "You can put together one of these little commentaries that you're always publishing in the ski hut."

February 23, 1952: The trend continues, and starting with the three hundredth generation the effects are as great as in the beginning. I really did set off an avalanche.

March 1, 1952: I proudly report my findings to Pauli who says, "But how did you make the individual effects finite, when they're in fact infinite?" I answer, "I used the regularization that you published with Villars." Pauli: "That was just a mathematical trick from us. Today it's known that this is the difference of large numbers, and if your results are to have any meaning in physics, you will need to estimate this difference." (In modern jargon, renormalization instead of regularization.)

March 2, 1952: I realize that the bounds for this difference will be an even more difficult problem, but after what I'd been through, I'm fearless.

April 1, 1952: Trash can filled with mountains of calculations, but no real progress.

May 1, 1952: I'm able to show that this difference isn't much smaller than my first result, which can therefore be kept.

May 2, 1952: Pauli enters the room, "You've gotten some competition. I'm getting an article from Angas Hurst, an Australian who's now working in England. He has made the same calculations and has arrived at the same conclusion. Take a look."

May 3, 1952: Hurst's article contained my first result, but not my others. Thanks to Pauli's criticism, I'm ahead by a nose. I can publish my article, and it doesn't have to join the mountains of calculations in the trash can.

P.S.: Angas and I later became good friends, and our articles—together with one from André Petermann, who independently came to the same conclusion at the same time—are still present in the literature.

P.P.S.: Even today, a half century later, it's not yet clear why electrons and photons don't have these avalanches of visitors. The fact that they could has been proven in numerous experiments and is undisputed. The creation of the universe, the big bang, was probably such an avalanche. Its particle flood discussed in chapter 1 was made up of virtual particles out of the underworld that had been set free by gravitational energy.

3

How Were the Chemical Elements Created?

IS MATTER BUILT LIKE A RUSSIAN NESTING DOLL?

Is the small always a repetition of the large?

AMONG THE GREATEST achievements of the ancient world were the atomistic teachings of Leucippus and Democritus. Through purely philosophical musings, they surmised a fundamental structure of matter; it would still take thousands of years for empirical confirmation. Their ideas weren't just some lucky guess in the realm of fantasy; rather they really touched on the core of the problem. They argued that matter is something that is indestructible, but can change in form; it has both of these properties and is made up of indestructible parts called atoms. Different combinations of atoms would provide matter with various forms. This is exactly how we think today—granted, our view of things is much more precise. It's only in the modern era that it has been possible to see an atom directly. Atomistic theory had been subject to debate over centuries, with the argument reaching its high point at the end of the nineteenth century. The atomists associated with Ludwig Boltzmann could explain a lot using the hypothesis of atoms—thermodynamics, for example—but the energeticists associated with Wilhelm Ostwald remained unconvinced; they believed in a continuous underground to all existence. When at the beginning of the twentieth century Wilson's cloud chamber made traces of separate atomic particles visible, the energeticists had to admit defeat. Boltzmann's ideas triumphed, but unfortunately only after his death. Today, when we ask who was right, the answer is no longer so simple.

For his own purposes, of course, Boltzmann was right: at the chemical level everything appears atomistic. At the deeper level of quantum field theory, the field serves as a continuous backdrop to everything that happens. Particles are just localized excitations; only under certain conditions do they last forever.

The materialistic-deterministic view of the world developed from atomism is no longer tenable today. As we've said, what happens in chaotic mechanical systems after a long period of time is not fully determined by the possible initial states. Quantum mechanics has produced a universal limit for this uncertainty of the state. I wouldn't like to use the word "causality" at this time. For some philosophers, it's a taboo; for many physicists, it's been refuted by quantum theory. This is subject to extensive argument depending on the exact meaning given to the simple statement, "Every effect must have a cause." The word "predictability" seems more appropriate to me, and is not unreservedly valid in modern physics. The belief that, in principle, even the remaining uncertainties can be determined by the movement of some hidden variables is an ideological superstructure that has no foundations in the known laws of nature. The position of the Aristotelian philosophers was the opposite, namely, that not only the local interactions of atoms determine what happens, but that there is also a striving toward an ultimate goal. This last idea corresponds to a principle in the fundamental equations of mechanics, but still the Aristotelian position can't be further developed to a productive theory of physics. Using a simple mechanical model, we will see in the next chapter how the terms "randomness," "necessity," and "ultimate goal" all play a role and are legitimate in their own right.

The first atoms found were themselves made up of parts, the electron 10^{-10} m across and the much smaller 10^{-14} m nucleus. The nucleus could also be divided into protons, P, and neutrons, N. The question was whether we have something like a Russian nesting doll in front of us. When you screw open one part, there's another smaller one inside. Is it possible to repeat this process forever, or do we at some point come to an end? The answer to this question is naturally colored by prejudices. Vladimir Lenin once said that matter is inexhaustible; therefore, his followers subscribe to the nesting-doll view of the world. On the other hand, Heisenberg thought that there must be an end at some point; preferably now as then he could "end" physics and develop the final theory in physics. Philosophizing itself couldn't provide an answer to the question of whether it's an infinite or finite nesting doll. The answer could come only by conducting the experiment itself. Today we have the answer, and it is: neither nor. As the German poet and satirist Wilhelm Busch said, "There is no doubt, here, as usual, things will turn out differently than you think."

I should praise Heisenberg a little more before I begin rejecting his final theory. When he was a young man of twenty-four, he had already written the most fantastic intellectual success story of the twentieth century. He had attempted to describe the movement of the electrons in the electron cloud in the usual way using data on their positions x and momenta p. He stumbled across one contradiction after the other, and, after numerous attempts in an almost mystical inspiration, finally postulated for these quantities the relation that now bears his name:

$$xp - px = i\hbar \text{ (H)}$$

$\hbar$ denotes the familiar Planck's constant, which is here spiced up by the imaginary unit i. He didn't care about what exactly that all meant; he simply used (H) and rushed along from success to success, including the Nobel Prize and the eternal fame in having been one of the founders of the deepest theory in physics, the quantum theory. What (H) means mathematically was a puzzling riddle, which would take time to unravel. Max Born, one of Heisenberg's mentors, was an educated man and knew that the commutative law for multiplication $xp = px$ was not valid for matrices. A matrix portrays operations like rotation, and multiplication means conducting operations one after the other. For example, the commutative law does not hold when you take a book and rotate it 90° on a horizontal axis and then 90° to the right on a vertical axis. When you then perform the rotations in the reverse order, the book will be in a different position. Therefore, it was thought that x and p were matrices and the whole thing was called matrix mechanics. It was soon discovered that (H) couldn't be fulfilled using matrices, and a retreat was made to a more abstract standpoint: x and p were operators. An operator is something that makes a vector out of another vector. It was found that (H) is only valid when x and p can change an arbitrarily small vector into an arbitrarily large vector. These kinds of operators are called unbounded. To save (H) it was postulated that x and p are unbounded. The problem with unbounded operators is that they can't be defined for all vectors, but only for certain vectors which form a so-called domain. (H) can only determine something when its defining domain is also given; there are different operators generated by different domains. Heisenberg never had an answer to this question. He wasn't even aware of there being a problem. This had no effect on the success of Heisenberg and his circle; they just took what they needed. Looking back, we have to give Heisenberg credit for having done the right thing; it would have taken years if he had stopped to try and decipher (H). Meanwhile Schrödinger's followers would have taken the cream of the crop. It was only decades later that physicists became more burdened by scruples, and to ease their conscience, an old branch of science, mathematical physics, was put on this new scent.

Figure 3.1: My first encounter with Heisenberg.

In 1950–1951 I worked with Heisenberg at the Max Planck Institute in Göttingen, and I learned a great deal from him. As the director of the institute, he rarely had time to discuss matters with us, but he regularly came to the physics colloquium once a week. There he joined in our thinking aloud when things got exciting. We were witness to how a genius actually thinks. For instance, once the discussion turned to photons, the light quanta, and Heisenberg philosophized aloud:

"A photon, how am I to imagine what a photon is? It's a particle that's always flying at the speed of light, so that the Lorenz contraction means that it's flat as a board. And how large is this little slice in diameter? Because it interacts with the electrons, its radius will be about as large as a Compton wavelength of the electron, 10^{-11} cm. [Heisenberg thought in centimeters, not meters.] But the cross section for scattering to electrons isn't 10^{-22}cm^2, but 10^{-26}cm^2, so it must be very transparent. Therefore, we should think of a photon as being a very thin, transparent, small slip, 10^{-11}cm wide." Of course, Heisenberg was breaking one of the commandments of quantum particles, "Thou shall not try to imagine what I look like." I was infected by this bad habit of thinking in mental pictures, and it has proven to have been quite useful.

Back to Heisenberg's last theory. A nuclear diameter of 10^{-15} m had been reached, and Heisenberg and several others thought that it was impossible to go further—or rather—smaller. Smaller lengths correspond to even greater energies, and Heisenberg went so far as to speak out against the development of even larger accelerators. He was of the opinion that nothing new would come out of it anyway, and everything up to 10^{-15} m would be described in his equation :

$$\delta\psi=\lambda\psi\psi^*\psi \qquad \text{(H II)}$$

However, things didn't turn out that way. Not only was it more difficult to make mathematical sense out of (H II) than it was out of (H): λ is an ordinary number, but the commutative law of multiplication wasn't supposed to apply to the symbol ψ, either; still, that could have been worked out somehow. Niels Bohr's objection that Heisenberg's theory was crazy, but not crazy enough, didn't get to the root of the problem. Heisenberg's last theory wasn't a success because the basic idea was incorrect. 10^{-15} m wasn't the end of the world, but rather the beginning of a new one that had up to then remained hidden. Today, it's been opened up to 10^{-18} m. It's populated by strange structures with unusual abilities, which make the alternatives of "finite or infinite nesting doll" meaningless. In the subnuclear world there are also particles with either tiny or enormous masses that are nonetheless intertwined through a confusing abundance of symmetry relations. Despite the large number of unexpected features, many distinctive structures were predicted before this empire was entered into. The human mind was able to predict nature's blueprint for areas that have taken us beyond the dark side of the moon.

THE STANDARD MODEL OF ELEMENTARY PARTICLES

The legions of elementary particles fit into a simple pattern.

It seems that we have gone a little astray in our line of argument. First we were using the entire universe as our basis and have wound up at the smallest bits of it. That these cornerstones of the very large and the very small are directly connected was first emphasized by the other founder of quantum mechanics, Erwin Schrödinger. Let me dig out a few more memories so I can introduce him as well.

Figure 3.2: Erwin Schrödinger

Schrödinger was a completely different type of person than Heisenberg. It was like he was always being chased: from one problem to another by his genius, from one country to another by the political powers in the twentieth century. He was a man full of contradictions. His original ambition was to hold a professorship at the provincial K.U.K. University in Czernowitz to be able to have the necessary peace and quiet to devote himself to philosophical investigations. This part of Austria was, in his words, "shot out from under his saddle" in World War I, and he became an internationally famous physicist. He demonstrated impressive civil courage after Hitler came to power in Germany and, as Max Planck's successor, left the most prominent German professorship in physics. As the Nazis caught up with him after occupying Austria, he was forced into a pathetic show of solidarity with the terror regime. Although he was known throughout the world, in exile he lived modestly with his wife in a small house in Clontarf, a suburb of Dublin. When I worked in Dublin in 1949–1950, he graciously invited me to stay with them at the beginning, although they didn't have any steady help. When I wanted to help him dry the dishes, he demurred, "No, no, it's no problem." Despite his charm, he looked rather surly on the—meanwhile obsolete—thousand-schilling note. He was a patriot, but only returned home as a burned-out old man.

In Dublin, he was still quite sociable, and always came to morning tea at the Institute of Advanced Studies to discuss all kinds of things with us. He liked to make fun of elementary particle theories and never mentioned the fact that he had just provided them with important impetus. Ten years earlier he had predicted, for example, the Hawking radiation we'll be discussing shortly in the de Sitter universe. He had always been fascinated by this type of space and wanted to know what it was like to live there, not how it came to be. He solved a wave equation in this space and found to his horror that every solution of a positive frequency also contained a part of negative frequency. To explain his negative reaction, I should mention the fact that, in quantum mechanics, solutions to wave equations from one frequency correspond to particles from this world, and, from the opposite frequency, to particles from the underworld. In the de Sitter universe, the particles from here are somehow attached to particles from beyond and pull them up into the daylight. Schrödinger was correct in indicating this to be a spontaneous creation of particles—or more correctly—in implying, as he didn't make too much of this. Only much later did Stephen Hawking recognize that the underworld shines like a light source at a certain temperature in a deep gravitational hole. However, the de Sitter universe isn't a hole; there isn't any outside. Still, Hawking demonstrated together with Gary Gibbons that this doesn't matter; there is Hawking radiation in the de Sitter universe, too. You could now conclude that we actually are living in a de Sitter universe and that the cosmic background radiation is actually this Hawking radiation. However, this can't be the case as the wavelengths of both these radiations are quite different. The Hawking radiation of a gravitational hole can be understood as being something like a gift certificate from Einstein that can buy particles in the underworld. It's only good for the cheapest—that is, the ones with the least energy. The energy of a photon is proportional to its frequency and therefore inversely proportional to its wavelength. Only the particles with the greatest wavelengths that still fit in the hole are available. The conditions are analogous in the de Sitter space; there the wavelength of the Hawking radiation is about equal to the radius of the space. The wavelength of the observed background radiation is the size of several millimeters; to paraphrase Nestroy, that's a completely different situation, and these cannot be the same radiation. If the de Sitter phase of the universe took place right at the beginning of Creation as the universe had the size of 10^{-34} m Planck's length, then its Hawking radiation would have the energy of Planck's mass. Then all gates to the underworld must have opened, and all types of particles would have been pouring out.

Whatever the details were, Schrödinger was right; the beginning of the cosmos was dominated by elementary particles. Although Schrödinger met with little response to this suggestion in his day, it is now the credo of particle physicists. We have yet to explore their world up to Planck's length, but we have been able to more or less reconstruct what happened after the universe swelled up to 10^{-18} m. Of course, we're not able to open the gates as wide as the big bang could, but it's possible to drill small holes to the underworld by accelerating particles to the energy corresponding to 10^{-18} m. This energy corresponds to about 100 proton masses. (Keep in mind the equivalence relation energy $[E] \leftrightarrow$ mass $[m] \leftrightarrow 1/$length $[1/\lambda]$ expressed as $E = mc^2 = \hbar c/\lambda$. As Planck's units, $\hbar$ and c have the numerical value of 1 so the relations become equalities.) Protons or electrons with this energy already have almost the speed of light, and if you let them directly crash into each other, you've produced a small big bang. Hundreds of particles jump out of the underworld, and if you make them visible, it looks like a jungle. Is it possible to decipher this chaos? Can order be seen in the background? Yes, and the order has been described in the so-called Standard Model, which we were justified in letting take over the title of this section. It predicts events with a margin of error of 0.1 percent, and the experiments support these predictions with the same degree of accuracy. That the human mind was able to work its way forward in these areas of the creation, which had lain dormant for billions of years and never been touched upon in biological evolution, is one of the most amazing achievements of the twentieth century. I would therefore like to sketch out this fascinating development.

Already in the 1930s, Heisenberg had noticed that some particles are actually siblings. P and N are almost the same particle; they differ in only one inner property, the so-called isospin. It points up in protons and down in neutrons. Up and down are, of course, just conventions as the inner space of the isospin has nothing to do with our space (or does it?); in any case it's three-dimensional. It's pretty amazing that our space is interwoven with another in which directions are marked globally. In this way a proton behind the moon also knows where up and down is in the isospace and where it needs to point its isospin. This sounds a bit peculiar, and that's why Chen Ning Yang, a member of a Chinese family of intellectuals who has been living in the United States since the end of World War II, developed a theory in which there's something like messenger fields that relay from space point to space point how they should orient themselves in the isospace. These are conventionally called "gauge fields," the accompanying particles "gauge particles," the symmetry in the isospace "gauge symmetry," and this theory "gauge theory." It took many years for this theory to be accepted, because a problem arose almost immediately. In 1954, when Yang introduced his ideas in a lecture in Princeton, Pauli was in the audience. I was also there and became witness to a remarkable debate. By nodding in his typical fashion, Pauli appeared to support Yang's remarks until the nodding turned into head shaking, and he jumped out of his seat.

Pauli: I must protest. Where are the particles belonging to your gauge fields? Have you already seen them?

Yang: No.

Pauli: They would have to have a mass of zero and be able to be produced easily. As no one has found them yet, your theory must be wrong.

Yang: I don't know. . . .

Pauli: That seems to me to be the main objection. It's completely irrelevant for us what the mathematicians are saying.

Yang: The question about the gauge particle's mass is a deep, dynamic problem, and I'm not in the position to answer it.

At this point Pauli wants to leave the auditorium and can only be placated by Oppenheimer, but he continues to shake his head throughout.

Yang's last remark was actually very farsighted. With what was known at that time, neither Yang nor Pauli were able to answer this question. Today, there's no longer any doubt that these gauge particles exist, but in theory, the problem of their mass has only partially been solved. In reality, some gauge particles are found to actually have a mass.

Pauli's outburst wasn't completely unfounded. At that time, there was already a prototype of a gauge theory, electrodynamics. The gauge field was the electromagnetic field, and the gauge particle fitting with it, the photon, had zero mass. It was one of the favorite dogmas of the theoreticians that gauge symmetry would force gauge particles to have zero mass. Dogmas had always annoyed Schrödinger, and already in 1951 he had construed a counterexample. However, he published it only in a small letter to the magazine *Nature*, which hardly anyone noticed, including Yang and Pauli. In 1961, Julian Schwinger gave a further counterexample. In his model, the space just had one dimension, and was therefore treated more like a curiosity. Only years later was the theory created that is now considered the cause of the mass for some gauge particles. The mechanism was suggested by Peter Higgs and others; Higgs had, without knowing Schrödinger's idea, further developed the same one. Many of the gauge particles found later (gluons) have the mass of 0. The reason that it was possible to see only the photon directly and not these gauge particles is something that no one at that time could have guessed.

It would still take a long time until the Standard Model came into being; new particles would be discovered, new symmetries found, but the principle of Yang's theory remained. All elementary particles with spin of 1 (intrinsic spin impulse $\hbar$) are gauge particles; all other elementary particles have a spin of 1/2 (intrinsic angular momentum $\hbar/2$), and their symmetries are conveyed by the gauge particles.

First let me explain a little about what "elementary particles" are. These are particles that are not made up of other elementary particles. Every particle receives guests from the underworld and is wearing a coat of virtual particles, and when the coat is taken off, a naked elementary particle

Figure 3.3: Sometimes Murray Gell-Mann needed more than words
to communicate.

should remain in the nucleus. The term "naked particle" bothered the
American physicist Geoffrey Chew; he thought that such a thing didn't
exist. He argued for nuclear democracy: everything was made up of everything. It was like in Edgar Allan Poe's "The Masque of the Red Death":
if you were to take off the mask, you'd find nothing underneath. Today,
it's an uncontested fact that every particle is wearing a coat of virtual particles from the underworld, and some have a thicker coat than others. It
is not a democracy; there are different ranks of particles. What was even
more surprising was that we know that some particles are compounds, but
we can't isolate the separate parts. It's as if we had a joke nesting doll:
when we shake it, we can hear that there's still something inside, but we
can't get it open. This is the case even with our old friend the proton. We
know that the proton is made up of three quarks, two u-quarks and one
d-quark. u and d stand for up and down and tell us the direction of the
isospin. To find out the deeper meaning of quark, you'd have to ask James
Joyce, as the name was borrowed from him. Murray Gell-Mann liked it,
and so it has stayed. Gauge particles called gluons tell us the directions in
the isospin area. Gluons glue three quarks onto each other. It would be
even better to call them supergluons, because they can actually do what
the makers of Superglue claim to be able to do. It's impossible to separate
the three quarks, regardless of which force is used. This phenomenon is
called "confinement," and it took a long time for physicists to get used to
it. This is the reason no one had even seen the gluons earlier, even though
they have a mass of zero. Gell-Mann, who suggested and christened the
quarks, originally thought that these particles were connected by huge

potential barriers. This revolutionary idea then made him a bit uneasy, and he decided to just give quarks a mathematical symbolic value. This idea was also suspicious to the positivists. I can still remember an argument with Heisenberg when I had spoken about the quark model's ability to predict at a conference:

Heisenberg: But where are your quarks; where do they exist?

Me: Well, in the proton.

Heisenberg: But you can't call that existing.

Me: At the beginning you also couldn't split up the atom, but you still believed in the atom model.

Heisenberg: But there at least you could see the electron; no one has seen a quark yet. I think that all of your successes with the quark model have just been coincidence. There's nothing real behind it all.

Meanwhile the question as to whether quarks "really" exist even if they can't be broken out of their chains has been handed over to the philosophers. The consensus is to think as if they exist.

Today, we have a pretty simple overview of the elementary construction blocks of matter. They are divided into two groups of spin 1/2 particles, the leptons and the quarks; and one group with spin 1, the corresponding gauge particles. That's it! Whatever else is crawling and flying through the particle jungle is made up of these.

The six leptons are written as:

$$e \qquad \mu \qquad \tau$$
$$\nu_e \qquad \nu_\mu \qquad \nu_\tau$$

and the six quarks are:

$$u \qquad c \qquad t$$
$$d \qquad s \qquad b$$

Similarities between the groups:

There are always pairs of spin 1/2 particles whose electric charge like P and N differ from the standard charge. There are three pairs in each group; they differ in their "flavor."

Every particle always has an antiparticle; e⁻ stands for the electron and e$^+$ for the positron. Both groups take part in and are connected by the electroweak interaction.

Differences between the groups:

1. While the top row of leptons has the electric charge of 1 (measured in elementary charges) and the bottom row 0, the top row of quarks has the charge of 2/3 and the bottom –1/3. (The antiparticles always have the opposite charge to the particles.)
2. Each of the six quarks is actually a triplet; it has two more identical brothers. This provides another occasion for great inner symmetry. The corresponding gauge particles are the gluons mentioned before. They are incredibly aggressive and prevent us from meeting quarks on their own.

Other than these spin 1/2 particles, there are also gauge bosons. There are four of these for electroweak interaction, and they are written as the letters γ, Z^0, W^+, and W^-. There are eight more gluons for the strong interactions between quarks.

The Standard Model boiled down to a manageable size what's fundamental in the enormous zoo of elementary particles. Nevertheless, it goes against the grain to see the deepest consituents of matter exhibiting such bizarre outgrowths, and the questions about what's behind the Standard Model have never been completely answered. There are many suggestions, but until now, the Standard Model has remained impervious to attack and has never shown any contradiction to experimental results. This model still can't explain everything, and the next point on our agenda is to discuss what questions remain open.

WHICH PARTICLES SURVIVE?

We see only the last fragments of the fundamental structures.

In case the reader has followed this cram session on elementary particle physics, the question comes to mind: "Who benefits from this? I admit that this landscape full of symmetries that are partially destroyed has a bizarre fascination, but isn't it like the world on a far-off star that in the end doesn't matter to us?" I have to take issue with this last bit. On the contrary, it's the basis for our lives. Then you could say that naturally we can once again explain why we can live, as in any case everything is controlled by the Standard Model. But as we will see, this isn't right either, as our existence is dependent on exactly those things which are not determined by the Standard Model, but just happen to appear there. I'm talking about the masses of the particles. I haven't mentioned them yet because they are not part of a success story in physics, but rather remain completely unintelligible. Their size appears to be randomly strewn across many powers of 10 (see Figure 3.4, page 66), and yet our curses and our blessings are determined by the fine details of this spectrum.

Let's begin with the leptons. The electron has a mass of approximately 1/2 MeV. MeV stands for a million electron volts, which is the amount of energy an electron would have if it were to fall through a voltage difference of a million volts. This voltage difference is achieved at a distance of around 10^{-15} m from an elementary charge, the electron charge. That's why the so-called classic electron radius of 10^{-15} m was considered to be significant. It turns out it isn't, because quantum theory already dominates at 10^{-13} m, the Compton wavelength of an electron, and prevents an exact localization. The electron charge would require the appropriate amount of energy for its mass if it were to be concentrated on 10^{-15} m, which quantum theory doesn't allow. So the electron's mass isn't really understood, even less the masses of its sisters μ and τ, which are 230 and 3,000 times as heavy, respectively. The electrically neutral siblings v_e, v_μ, and v_τ have a tiny mass; the amount is not exactly known, but could be around 10^{-8} electron masses. The conditions of the corresponding gauge particles are even more extreme.

The only way leptons communicate is by electroweak interactions (which we will be discussing directly), which have a four-dimensional symmetry; therefore this is where the four gauge particles mentioned earlier are found. One of these is our old friend the photon γ. It's always said that the photon has zero mass. Perhaps we should be a bit more cautious and say that the mass lies below any measurable size. If the mass is zero, the Compton wavelength becomes infinite; but if the radius of the entire world is only finite, then we can start messing around a bit. In any case, the mass of its three siblings Z^0, W^+, and W^- is incomparably greater— 150,000 times as heavy as an electron and 80 times a heavy as a proton. Though elementary particles, these gauge particles are as heavy as a medium atomic nucleus. The realization that γ, Z^0, and $W^\pm$ are related was the last great discovery in elementary particle physics. I'd like to devote a few lines to the story of how this was done, but mentioning all of the many protagonists would take pages and pages, so I can only more or less randomly pick out two.

First I would like to create a small monument for my old friend Bruno Touschek. He was a physicist from Vienna, who went to Germany in 1940, because, being half-Jewish, it was easier for him there. Nonetheless, he was deported to a concentration camp toward the end of the war. On the way he collapsed and was therefore shot and left for dead. He fell into the gutter, and the convoy continued on. He had been shot in the head, but hadn't died. Bruno was taken in by someone and nursed back to health. After the war, he could work under Heisenberg, who inspired him with the idea that the fundamental forces should be combined. Once in the early 1960s, he was in Vienna with me and said that now, like nineteenth-century politicians debating the creation of a nation-state, we've got to decide if we want the "small German" or the "greater German" solution. I'm for the first, because it's impossible to have everything at once." What he meant was whether we should first join the electric forces with the weak to get what we now call electroweak interaction, or immediately to reduce with the nuclear forces all forces to a single force. His preference for the "small German" solution has proven to have been wise, as the "greater German" has yet to have been successful. Touschek also had problems with the "small German" solution, but he forged the weapons to be used in its experimental verification. He recognized that the necessary energies could only be achieved by having a particle ray beat down on one running opposite, preferably an electron ray against a positron ray. He began to realize this principle in Frascati near Rome, where he was working at the time, first on a smaller machine called Ada and then with an enlarged machine called Adone. The realization was unfortunately delayed by the strikes in Italy in 1968, so that some achievements were first made by Gersh Itskovich Budker in Novosibirsk, who didn't have these problems. It had been proven that a noticeable collision rate can be had with two rays running against each other whose particle density is less than the vacuum density of a lightbulb. This principle celebrated its triumph with the LEP (large electron positron collider) at CERN. It was really large, with a circumference of 30 km, and so energy-rich that it was possible not only to prove $e^+ + e^- \rightarrow Z^0$, but also to measure it exactly. It was even possible to provide evidence that there are only three kinds of neutrinos, because another v_4 would have been noticeable through the decay $Z^0 \rightarrow v_4 + \text{anti-}v_4$. Unfortunately, Bruno didn't live to see this success.

Bruno was fascinated by many fundamental questions—for example, by the idea of going back in time. The point is whether what's happening in a film played backwards appears to be only macroscopically impossible, but remains tolerated by the microscopic laws. Bruno's broodings on these topics even brought him into conflict with the authorities because he had a penchant for two other things: Italian red wine and motorcycles. Taken all together this was an accident waiting to happen, and as he was pulled over, the police noticed his accent when speaking Italian. He was asked what he was doing in Italy, and he replied that he was working. The police officer wanted to have more details about what kind of work he was doing, and Bruno realized that here he had to stick to the truth. Bruno replied that he was working on going back in time, and the officer decided that the man was not only drunk but also insane, and Bruno was taken to a psychiatric clinic. Luckily, just then Valentin Braitenberg, a neurologist from southern Tyrol, happened to be on duty. He recognized that the apparent insanity was really genius. Bruno was released, and southern Tyrol became his favorite place.

The "small German" solution was finally achieved by, among others, Abdul Salam, called Abdus, an interesting man from Pakistan. He was educated there in a provincial school and liked to tell the following anecdote: Once in physics class they were discussing the different forces, and the professor listed them: "Friction, centrifugal force, gravity, and electrical forces. There are also nuclear forces, but only in Western industrialized countries." Abdus was a faithful Muslim; he believed in only one God. The idea that there were forces that existed only in some nations and not others was as unbearable as the idea that there were national gods. Therefore he devoted the rest of his life to unifying the forces of nature. He didn't succeed with the "greater German" solution, the unification of all forces, but considering all the difficulties that he had to overcome first, his achievements are remarkable, particularly the joining of electrical and weak forces. At first it seemed hopeless to unite the electric and weak forces; their appearances were too different. Weak forces cause the so-called β decay, for which $N \rightarrow P + e^- + v$ is a prototype. The force that causes this is so weak that a free neutron has a lifespan of about fifteen minutes. Only in the nucleus will it stay together longer due to nuclear forces. Analogous processes with photons are faster by twenty powers of 10. How can there be a similarity? The idea was that it was a two-step process; first $N \rightarrow P + W^-$, then $W^- \rightarrow e^- + v$. W^- is so heavy that it can only be generated "virtually" or by taking out an energy loan. That costs a huge Gamow factor, which makes the β decay forces seem so weak to us. When they are initially as strong as the electric, then we can predict how heavy the W-particle has to be to weaken them enough. Their mass was huge for that time, as we've said—like an average atomic nucleus. Salam had to wait more than twenty years before acceleration technology was developed sufficiently to be able to create W- and Z-particles. Abdus lived to see this, and the Nobel committee didn't hesitate and awarded him the Nobel Prize shortly beforehand. Then his powers were extinguished, as he was afflicted by a rare nerve disease. The last time we were together I had to help him up from his chair. It wasn't his fate to live with such a disease, just like Stephen Hawking.

Salam still wasn't able to say which mechanism created the huge mass for $W^{\pm}$ and Z^0. Peter Higgs then postulated that there were particles that could do this; the search has been on for these Higgs particles ever since, up to now unsuccessfully. The search conditions will improve after 2007, when the new super accelerator is working at CERN. Until then, I will have to leave this part of the story open-ended (perhaps then this book will be reissued and I really will be able to come to an end).

I can hear the attentive reader ask, "And what do I get out of all of that?" and the expert in elementary particle physics will answer, "From my point of view I don't see any practical applications." Here I would repeat the definition of an expert, from David Hilbert: "An expert is someone whose mental horizon has shrunk to a point and calls this their point of view." If we include other specialties like astrophysics (which we will touch on in the next chapter), then we'll see that this is all vitally important for us. If $W^{\pm}$ and Z^0 had zero mass like a good gauge particle should, then it would pass on the same strength of interaction with matter to a neutrino as to its leptonic sister the electron, and it would never emerge from inside the star. In this case, the entire supernova mechanism couldn't function, and the carbon and calcium needed for our bones could never break free of the star's center. There are many other examples, like when the neutrino mass is too large, the virtual process P → N + e⁺ + ν would cost too much energy, so that after the big bang, we wouldn't even make it to deuterium. The most amazing life insurance we find directly with the quark mass spectrum.

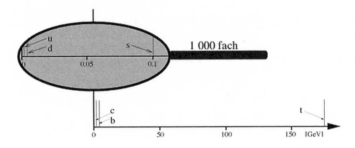

Figure 3.4: Spectrum of quark masses in GeV units. For the light u, d, and s quarks, the scale has been extended by a thousand times.

The mass of every quark is equal to that of its anti-quark, but otherwise the mass values of the six quarks cover a wide range. While m_d is only a few electron masses greater than m_u, m_s is a few hundred higher and m_t is already thousands electron masses higher. The consequence of this is that all agglomerates that can be built out of quarks are unstable, and they decompose into lighter combinations. The only exception is the proton which is built up out of (u, u, d). Its ongoing stability is also occasionally called into question, but it's stable enough for our purposes. The structure is made a little more complicated by the fact that their every quark comes in three designs, the so-called colors. The term "color" is somewhat misleading because it means something like an electric charge and can have a positive and negative value. We call the negative of a color simply the opposite color, much like normal colors have complementary colors, e.g., cyan is the complementary color to red. The thing to remember is that particles with opposite colors attract each other with such a powerful force that only agglomerates with a total color of zero can walk around unmolested. The easiest way to neutralize color is when a quark joins an anti-quark, since they have the opposite color. There are 6 × 6 = 36 possible weddings, and the married couples are called mesons. Some have already celebrated their golden anniversary, meaning that they were discovered over half a century ago, namely the π-mesons (where ū means an anti-u-quark):

$\pi^+ = (u, \bar{d})$, $\pi^- = (d, \bar{u})$, $\pi^0 = $ a mix of $(u, \bar{u})$ and $(d, \bar{d})$.

These are the lightest. If you replace u or d with s, then the mass is immediately increased by a factor of three and you get the K-mesons which were discovered later,

$K^+ = (u, \bar{s})$, $K^- = (s, \bar{u})$, $K^0 = (d, \bar{s})$, $K^0 = (s, \bar{d})$.

In the upper right hand corner we have indicated the electric charge of each meson. Let's check and see how the total charge ends up as an integer:

π-mesons: $2/3 + 1/3 = 1$, $-1/3 - 2/3 = -1$, $2/3 - 2/3$ or $-1/3 + 1/3 = 0$,

K-mesons: $2/3 + 1/3 = 1$, $-1/3 - 2/3 = -1$, $-1/3 + 1/3$ or $-1/3 + 1/3 = 0$.

You can continue to create new mesons in this way by using all thirty-six quark/anti-quark combinations. These present many researchers with a nice playground, but have little bearing on everyday life; they all have an extremely short lifespan. When we go up the list, the quark masses increase dramatically. Already K-mesons have the decay $K \rightarrow 3\pi$. It doesn't upset the conservation of energy. But what about the lightest meson? It decomposes, too, but only because the weak interaction connects quarks with leptons. The decay goes like $\pi \rightarrow e + v$, but is less frequent than $\pi \rightarrow \mu + v$, which is also energetically possible. A wise guy could now speak up and say, "That's only because the π-meson weighs the almost obscene amount of 300 electron masses. The masses of u and d are much smaller. Why isn't it satisfied with a mass so small that not even $\pi \rightarrow e + v$ is possible?" I can only reply that it might be so heavy because there are many more gluons hanging around besides the quarks. But no one knows for sure; some people have a completely different idea. Many even envision a world where the π-meson has zero mass. Now the wise guy can smirk, "But that would be fun as then the π^- meson would outdo the electron. Not only could it be stable, it would even destabilize the electron because $e \rightarrow \pi + v$ would be possible." That would be too much for me. I would have to say that that wouldn't be fun at all, but would be a catastrophe! Wise guy: "Why are you getting so angry. Instead of e, you would anyway have π^-." I'll now explain to the reader in more detail why this is not a comforting thought.

The two classes of elementary particles we use in the Standard Model differ from one another in an important property: the particles with spin 1 are called bosons after their discoverer Satyendra Nath Bose; those with spin ½ are called fermions after Enrico Fermi. Just like with even and odd numbers: fermion + fermion = boson, fermion + boson = fermion, boson + boson = boson, so you can say that one quark is a fermion, two quarks together are a boson, three quarks are a fermion again, and so on. The lifestyles of the two kinds of particles are completely different: bosons are bourgeois and fermions are aristocratic, meaning that the bosons like to remain together, but the fermions and therefore also electrons distance themselves from one another. The aristocratic nature of the electron lends the matter the following structure:

1. When the electron cloud is so complete so that the nucleus is neutralized, it will take in one more electron at most. There are, therefore, only simple negatively charged ions.

2. In the periodic system of elements, you're always running into noble gases, whose atoms don't want to have anything to do with each other.

3. In matter, the atoms are quite chaste in their contact with each other, rather than melting into each other.

This all would change when the π-mesons make them middle class. The atomic shell of neutral atoms would take in up to 20 percent more π^-; all atoms would resemble a standard atom and would greedily join up into molecules; the more matter they grabbed, the more these clumps would shrink. There wouldn't be any energy problems in such a world. After each gluttonous feast, the clumps would get hotter, but relatively modest creatures like us couldn't survive there. Here comes our wise guy again: "I don't want to mess with your beloved electrons any more, but you've got to admit that the π-meson with its mass is so different from its meson siblings that it can only be caprice on its part. It could have anywhere between 10 and 1,000 electron masses, μ-decay is ultimately unnecessary." I have to remain firm: I can't barter anything from my π-meson mass m_π; the architect calculated here exactly with no room for leeway. This mass is determined by the Compton wavelength of the π-meson, 10^{-15} m, and this in turn tells us the reach of the nuclear forces. It has been doled out precisely so that the deuteron D = P + N stays together, but super light helium ^{2}He = P + P doesn't. This is balancing on the edge: the negative energy of nuclear forces opposes the positive kinetic energy of P and N. The kinetic energy, according to quantum theory, is produced when they thrash about wildly against being locked up so tightly. This difference is only a small fraction of the entire "thrash" energy, and if I were to double m_π, it would even quadruple. Even a small increase in m_π causes the destruction of D; a small decrease already produces ^{2}He. Both would ruin the staging of the big bang. If there were no D, there wouldn't be any opening to construct heavy elements, and we would be living in a pure hydrogen world. If there were to be ^{2}He, hydrogen would immediately burn up into helium because P + P $\rightarrow$ ^{2}He + γ would happen very fast and we would be living in a pure helium world. We wouldn't be able to exist in either case.

Let's shake off these nightmarish visions and get back to our quarks, which are indeed connected, but still aristocratic. If we want to get by without anti-quarks, then it is already more difficult to produce colorless combinations, and these are the only kinds that can freely exist. Every quark has color, but it doesn't work with two, either, just like two quarks together can never have an electric charge of zero:

$$2/3 + 2/3 = 4/3,\ 2/3 - 1/3 = 1/3,\ -1/3 - 1/3 = -2/3.$$

With three it's like with the charge:

$$2/3 - 1/3 - 1/3 = 0.$$

The three-quark-systems are called hadrons and form a colorful menagerie of $6 \times 6 \times 6 = 216$ kinds. Here we run into our old friends P = (u, u, d), N = (u, d, d), and also historically important companions like $\Omega^- =$ (s, s, s), whose discovery lent credibility to the entire system. Two hundred fifteen of the possible combinations are quick to decay, but there has to be one remaining, as the number of quarks seems to have been maintained very well and hadrons can't dissolve in leptons. On the other hand, hadron $\rightarrow$ hadron + meson works, because fermion $\rightarrow$ fermion + boson is allowed. The lightest, therefore, survived, and that is the proton. This is connected to the fact that the u-quark is the lightest, but $N^{*+} =$ (u, u, u) is already heavier. Their aristocratic nature prevents three u quarks from snuggling up like this.

Once again we've been lucky. It doesn't bear thinking about what would happen when the neutron—and not the proton—were stable. Then there wouldn't even be hydrogen H = P + e⁻. The world would just be made up of neutrons that would cluster together into neutron stars that could only offer a dark stage. Richard Feynman was the first to point out this dark possibility and to look for a possible explanation. It should really be just the opposite; the electric charge makes a particle heavier: $\pi^\pm$ is heavier than π^0, but $K^\pm$ is not heavier than K^0 or $\bar{K}^0$. The magnetic moment still comes into play, which is already larger for a proton than for a neutron, so that the magnetic energy should also make the proton heavier. Feynman discovered an interference effect between electric and magnetic forces that he could manipulate so that the neutron became the heavier of the two. He was very proud of this, but what he couldn't know was that he had only deluded himself. As we have seen, the root of the problem is found at the quark level: the d-quark is heavier than the u-quark and the neutron contains more d-quarks than the proton. But does this really explain why the proton is lighter? Absolutely not. We've just transferred the problem from between some rocks into a hard place. The u-quark has a charge of 2/3, the d-quark has −1/3, so that when things really get going, u has to be heavier than d, and we're faced with the same problem as before.

The popular Austrian writer Friedrich Torberg had his quintessential character Aunt Jolesch say: "God save us from our good luck." We will ask: the u-quark is admittedly the lightest. Is it then set in stone that the world can give birth to life? The answer is no, and once again we've been lucky. If d were much heavier than u, it would be deadly. Then the neutron would be so unstable that not even the nuclear forces could keep it together. There are no other atomic nuclei made up of only protons besides P; in this scenario, the only chemical element would be hydrogen. This world wouldn't be as dark as the last one, but there still wouldn't be any life in it.

The question now arises if all of these fortunate turns of events have come from our box of knowledge labeled coincidence or necessity. To put it dryly: are they dependent only on the dynamic basic law or also from the state they're in? The first certainly determines the spectra of elementary particles. The particles directly relevant for us aren't elementary; they're more like waves in the ocean. Their properties are less dependent on the forces between the protons and electrons and the atomic components of their matter than on local conditions. We can put away important characteristics of our particles in the box marked coincidence and uninhibitedly speak of a miracle.

The quark world tells us the fact that we exist is a (miracle)^5, a fivefold miracle:

Miracle I: d is heavier than u.

Miracle II: d is only slightly heavier than u, by not as much as the differences between other quark masses.

Miracle III: W$^{\pm}$ and Z^0 have gigantic masses so that weak interaction is sufficiently weak, but not too much so.

Miracle IV: The mass of π is so ingenious that D binds, but ^{2}He doesn't.

Miracle V: The electron masses are such that the process e$^-$ → π^- + ν and P + e$^-$ → N + ν are energetically impossible.

We have no explanation for any of this. It now occurs to me that I've made a faux pas. I've already quoted a German-speaking poet, but not yet the great Johann Wolfgang von Goethe. To rectify matters:

> *The greatest happiness of the thinking man is to have*
> *fathomed the fathomable, and to quietly revere*
> *the unfathomable. (Goethe)*

THE INNER SPACE

The most important things happen in a secret space.

The symmetry of the inner space dictates the dynamics in the Standard Model. Think back to what we said in chapter 3, that the spin-1 particles (gauge bosons) determine the directions in the inner space in every point of our space, and the way they hook up to spin 1/2-particles ensures that they (the spin-1/2 particles) don't lose their sense of orientation in the inner space. It just so happens that these factors limit the mathematical description to a large degree and determine the general structure of the motion equations. Only the strength of the coupling is taken from the experiment.

The ideas behind these gauge theories are very simple but involve incredibly complex calculations. These are the nonlinear equations, and, as in many areas, a modern mainframe is needed to get a numerical result. As the theory describes the particle's behavior with a precision of one part in

1000, these tiny bits act as highly precise analog computers for gauge-theory equations. Although we have come a long way in the miniaturization of computers, the idea that these complicated equations could be solved by a 10^{-16} m thing in 10^{-26} s—a collision only takes this long—appears incredible. Many people have no problem accepting this—they say that if there's a problem with it, then it's a psychological rather than physical one. However, I don't want just to drop this question, but would like to discuss a further point of vew that will make these mini-supercomputers more comprehensible. The evidence of such ideas reaches far back into the past.

To be brief, the idea is that it's just from our human perspective that elementary particles appear to be so small. In reality, they are quite complex structures, because the correct unit of length, is Planck's length and in these they have a size of 10^{19}. Planck's length is also the relevant size, because the inner space is ultimately also a component of the space in which we live; it's just that this direction didn't participate in the cosmic expansion and shrank to a Planck's length. That's where the effects of quantum gravitation stabilized this dimension, but the details as to how are not understood to this day.

The roots of these ideas can be found in ancient physics, when almost nothing was understood of quantum theory. We will trace this intellectual journey. In 1916, Einstein put the finishing flourish on his theory of relativity by developing the equations that rule over the gravitational field. However, David Hilbert was a little more brilliant in mastering the necessary calculation technique and published the equations before Einstein did. Einstein had provided the mental groundwork, so that they are justly called the Einstein equations. This completely new theory was too much for the scientific community, where reactions ranged from being insulting to mindlessly parroting everything. However, the effects prophesized as changes in the Newtonian gravitation theory were so tiny that the whole thing only remained a stumbling block for philosophers for a long time. Today, the global community enjoys a considerable benefit from the theory of relativity in the form of GPS (global positioning systems). Radio signals can be received from every satellite in space, and from the differences in the reception times, the position of the receiver can be reconstructed within 10 m. Many people benefit from this: ships at night or in fog, researchers in the jungle, lost visitors to the big city, and others. To achieve this fantastic degree of precision, even the smallest corrections must be taken into account—the same as the theory of relativity. Without it, GPS would be useless.

One reaction to Einstein's work pointed to future developments. In 1919, Theodor Kaluza offered an original extension to the Einstein equation to the Prussian Academy of Sciences for publication. He assumed that our space would have four dimensions, and space-time have one more—that is, five—but the reason that the fourth dimension isn't seen is only because nothing depends on it. He found that in this five-dimensional world the Einstein equations not only describe just the gravitational field, but that it's coupled to an electromagnetic field. In this peculiar way, gravitation and electricity are forged together perfectly. The expert to review his work at the academy was none other than Einstein himself. He must have felt a bit like Kaluza was trying to pull his leg with a fourth dimension that no one can see. Still, he apparently thought that something must be behind it and didn't reject this study. What he did do was absolutely nothing. He left it lying on his desk for two years; maybe he just kept putting it off. Finally he did accept it, but without first deciding if it made sense or not. We can't hold this against him, because up to today we still don't know. This study did introduce a paradigm, which many generations of physicists have since lived off. There is, namely, a simple answer to the question as to why we can't see the fourth dimension; it's shrunk! Imagine we were to live in a four-dimensional tube, whose three-dimensional axes represent our familiar space and the fourth dimension, the tube circumference, was infinitesimal. Nothing more can then develop in this direction, and the tubes would seem like a three-dimensional space. If you want to derive the strength of the electrical interaction from the gravitational interaction, then the tube radius must be about as small as a Planck's length. Since then, a lot of work has gone into trying to develop the inside of the tube to the inner space of the elementary particles, without changing their minute size. The question arises as to how our three spatial dimensions, which cover immeasurable areas, have such tiny brothers and sisters. The thing is, our space also began very small, and it's conceivable that other directions have already collapsed or haven't even expanded. As Planck's length plays a role, it must be an effect of quantum theory. This prohibits electrons from falling into atomic nuclei, so maybe it would also not allow the inner space to shrink to a point and would stabilize it at a Planck's length. To push forward in this jungle with an experiment, we would have to accelerate particles until their energy corresponds to a Planck's mass, which we're still far from being able to do. Right now, physicists can continue to elaborate on their tall tales— and there are certainly no shortage of those—in peace. For example, right now the "string theory" is being pursued, which presents particles as strings with a diameter of a Planck's length. Many people find the idea quite tasty that the ur-soup, simmering after the big bang, was noodle soup. However, there is no experimental basis for this theory, and no prospects for any anytime soon. Nevertheless, from today's point of view, it's unavoidable that the ur-backdrop of events took place in areas of Planck's length with energy of Planck's energy. The differences in mass among the elementary particles look like spiderwebs on a skyscraper next to these enormous energies. These webs have been so finely spun that they carry our lives; we are now entering curious worlds by digging deeper and deeper.

4

Do You Know
How Many Stars Are in the Sky?

A STAR IS BORN

Stars also experience birth, life, and death.

THE TITLE of this chapter is taken from a German children's song which asks, "Do you know how many stars are in the sky?" only to take a surprising turn and continue, "Do you know how many clouds are drifting around the world?" We've gone from one extreme to the other; stars are points of light in the sky that can be individually counted. They move according to the laws of nature and represent eternity. Clouds are the exact opposite. They are random formations that come and go and even flow into each other. They symbolize transience. The apparently naïve song then goes on to make an even more astounding point, "The Lord God has counted them all to make sure they're all there." The song is a perfect setup to the leitmotif of this book: there are things whose development is determined by the laws of nature, there are things whose fate is beyond the grasp of human reasoning, and both are subject to a higher power. I can already hear a physicist objecting: "That song is just filled with childish nonsense. There's also not that big of a difference between stars and clouds. In fact, stars are really only cosmic gas clouds that have pulled together. This has all happened in accordance with the laws of mechanics and Newtonian gravitational theory. There's nothing mysterious about it, so let's just stick to scientific facts!" That's exactly what Heide Narnhofer, a mathematical physicist; Harald Posch, an experimental physicist; and I set out to do, but in the present era of high-speed computers. The equations in Newtonian mechanics may explain how all those particles come together to form a star under the influence of attracting forces, but they are so complicated that not much information can be inferred when the calculations are made by hand. Modern computers, however, are another story, and by reconstructing a conversation from memory, I would like to give the reader an idea of our work, which new perspectives have opened up to curious physicists, and how the children's song has, in the end, been granted a deeper meaning.

Act 0

Me: What would be a good model to use to study how particles condense during mutual gravitational attraction?

Harald: To make sure we don't just get a confusing swarm on the screen, there shouldn't be too many.

Heide: But if there are not enough, we might not get a realistic representation of the situation. Is four hundred an acceptable compromise?

Harald: That'll be no problem for the computer.

Me: One difficulty appears to me to be that one particle after the other will take off, and we won't have enough left over. We should do it as if they are moving around in a closed universe. When a particle tries to get out of one side, then it will come in again on the other.

Harald: Since we've already started simplifying things, we should lessen the Newtonian attraction at small distances between the particles, because otherwise the motion is too fast for the computer to break it down.

Heide: Then let the force drop sharply at a greater distance. Otherwise we'll get long-term global movements that would cover up what we're really interested in.

Me: Now I've lost all scruples and would like to limit the motion to a plane. On the screen we're seeing everything projected to one level anyway, and there are also many flat objects in the sky. We'll accept all of these simplifications and comfort ourselves with the thought that it should only be a simple model that demonstrates the most important properties.

Act I

Harald is a computer expert who has already calculated the most complicated molecular movements. For him, four hundred particles with a comfortable amount of attraction on a torus are "peanuts," and he immediately developed the computer program. He wants to show us the resulting motion, and, full of anticipation, we're sitting in front of the computer, staring at the screen.

Me: Right now it looks like a swarm of bees to me—it doesn't really tell me anything.

Harald: There's only been a little bit of collision time. The computer divides every impact into thousands of steps, and for every step it has to calculate the changes in speed and location for all four hundred particles. In this short time it has already completed millions of calculations.

Heide: Now lots of particles are gathering in a heap, as if they're gathering around a queen.

Me: There are no queens here, all particles are equal. The heap has already dissolved. But now they seem to be rushing toward a center.

Heide: There isn't a center, either; all points are equal. Now they are forming several strange-looking dumplings.

Harald: Now one clump of several particles has blown up a smaller one and swallowed up the pieces. It's like in an aquarium where the big fish eat the smaller ones.

Me: Finally we have just two aggregates; all the rest have already been caught. I think we can go ahead and call these stars.

Heide: These two aren't letting anything else come up. They're only just drifting around.

Me: Yes, now it's starting to get boring already. Only a cosmic catastrophe (when they collide) can liven up the monotony.

Harald: They finally look like they're on a collision course, but then they decided differently.

Me: Who would come out the winner after they collide? They're about the same size.

Harald: Neither one would win. Both would lose a few feathers and a giant star would be created, which would then be the sole ruler.

Heide: Now it's getting exciting—they can't avoid each other anymore.

Me: There we have the "big splash." As you said, Harald, there's one star left that's larger than its predecessors, but lots of particles were also blown into the atmosphere. Of course, in all the excitement I was talking nonsense; it's impossible to say who won.

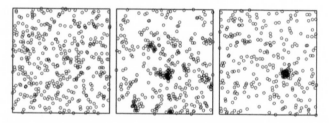

Figure 4.1: Two-dimensional model for the birth of a particle cluster.
The three snapshots show, at left, the initial state, a random gas of particles;
center, an intermediate state with several small clusters shortly after
the start of the simulation; and, right, the final state close
to thermodynamic equilibrium: one cluster is dominant.

Heide: This one star also seems to control everything; we have something like the Pax Romana. How can that ever change?

Me: I don't think it can. But I'd like to look at two things in more detail. In all the tumult, it's impossible to follow the fate of a particular particle. Are they always caught in a star, or are they sometimes able to escape? Second, it seems to me that the particles in the star are much faster than they are outside. Things are whirling around in there so much that you can't be sure.

Harald: I can answer both questions by using color. I'll give one particle a different color from the other 399, then it will stick out and we can see what it does. I'll also make the slower particles blue and the faster ones red, and then we can immediately see where the hot spots are in our sky.

Act II

Harald has introduced color; Heide and I have discussed some more theoretical considerations, and we're once again sitting in front of the computer.

Harald: I just took a random particle and colored it as a small white ring. I think it sticks out quite distinctly from the rest.

Me: Our test particles appear to be shoved around aimlessly. I can't see any regularity.

Heide: Yes, it's a real random walk. That's what Brownian motion is supposed to look like.

Harald: Now the test particle is getting close to the star. I'm curious to see if it's going to get caught.

Heide: No, it just bumped into it but escaped on the other side. So the star isn't a perfect vacuum cleaner.

Harald: Now the test particle doesn't seem to even notice the star anymore.

Me: But now it has happened—it has been caught by its toe.

Heide: Yes, and it's squirming around inside the star, which shows how hot it is inside.

Me: It reminds me of *a*-decay, where the *a*-particle in the nucleus also has to make many escape attempts before it's successful.

Harald: Now it's time for the big splash where the stars are going to crash together. What will happen to our prisoner when that happens?

Me: Whoosh—it has gotten outside again. It used the opportunity to escape. There are not a lot of new things that can happen from now on. Speaking of hot, let's examine the temperature conditions more closely.

Harald: The stars are red, the atmosphere is blue, so that it's obviously much hotter inside than outside.

Heide: I don't like that because Boltzmann drummed it into us that the universe has to one day perish by heat death. In that case, it's got to be just as hot outside as it is inside.

Me: Maybe we haven't waited long enough. When some hot particles escape and cold ones get captured, then the temperature has to even out at some point.

Heide: I'm no longer so sure about that. It's so hot inside just because the particles heat up when they crash in. We're not witnessing anything other than a cosmic sirocco.

Harald: The best thing for me to do is to use a quick-motion function. I'll skip several thousand impact cycles, the computer can calculate these faster than we can see them anyway, and then we'll be able to see if the whole thing has heated up.

Heide: Everything really is uniformly red. Boltzmann was actually right about heat death.

Me: Still, it seems to me that the initial situation of imbalance is more important for us. We live off sunshine, and that exists only because there's a difference in temperature that has built up between the sun and the earth. If we were to have the same temperature, then we would radiate back exactly the same amount of energy to the sun as we receive from it. But we've gotten a short-lived reprieve from Boltzmann's heat death.

Harald: Before we venture so far into the future, let me make sure that the computer program will continue to function reliably for such a long time.

Act III

We're sitting in front of the computer again.

Harald: My computations don't provide a conclusive answer for such a long period of time, even though I'm calculating to a precision of sixteen figures. I can tell, because after a star has built up, I suddenly turn the speed of all particles around. Then, the motion should have to go backwards like a film on rewind, and the star should dissolve again. When I record the number of particles in the star as a function of time, then at first this number acts as expected and meekly follows this number during the building of the star. It even copies all the little fluctuations. After a while, though, the accounting gets out of rhythm, and then the reversibility is lost. Instead of dissolving the star in its entirety, it becomes larger again after some time, and at the zero hour there's a star although there wasn't one there at the beginning (Figure 4.2). I can only explain this by the unavoidable calculation errors the computer makes with endless decimal places, which then simulate this effect.

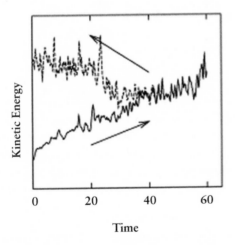

Figure 4.2: Temporal development of kinetic energy, which is a measure for the size of the cluster built. At time t=60 (in dimensionless units) the speeds of all particles will be reversed. The following motion corresponds to a time reversal, which is represented by the upper arrow.

Heide: That means that the errors from rounding off transfer the trajectory to one lying close by, but over time it leads to somewhere else entirely. We can test this by using two different, neighboring starting points for the trajectories from the very beginning. After some time, they will have to show a marked separation and lead to completely different things.

Me: We should define the closest neighbors as having the same total energy and momentum. Let's just exchange the initial speeds of two particles far away from each other that have nothing to do with each other. That wouldn't change the total energy or the total momentum and should produce the same state as far as it's humanly possible to tell.

Harald: I can do that easily; it would be best first to change just one component of their speed and then the other.
The computer hums along for a while, and then what we've been anxiously awaiting appears onscreen (Figure 4.3).

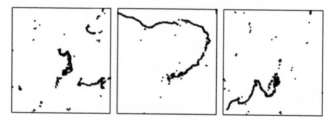

Figure 4.3: Movement of the center of mass for the largest cluster at the time. At left: a cluster orbit. Center and right: the initial configuration has been slightly altered by exchanging the x and y coordinates of two particles located far away from each other.

Harald: You see, sooner or later a star gets built in all three cases, but where this happens and where it then drifts off to are always different. When we compare the trajectories from the three case scenarios, they show absolutely no similarities.

Heide: That's just what we would expect according to our theoretical studies. We looked at what the state space has for structures at the total energy used. We found that the greatest volume by far was shown in structures with a star in the state space. Therefore, with overwhelming probability, almost every trajectory has to pass an area like this, and a star will be formed. However, in our geometry, every point is equal, and the probability of a star being created just there has to be the same at every point. Apparently, the trajectories where stars form at different points are so interwoven that they can hardly be separated using the starting points, so it's impossible to plan at a certain point at a particular time.

Harald: I'm still uneasy by the idea that the rounding-off errors end up by showing a scenario that isn't true. I will study the same question on a discrete system where the computer only uses whole numbers and so there won't be any rounding-off errors. Then there won't be a random element mixed into the calculations, and everything can then only be caused by the sensitive dependence on the initial conditions.

EPILOGUE

We're relaxing around a cup of coffee and chatting about our computer adventure.

Harald: The discrete system showed the same results as the continuous. I replaced the continuous speed of the particle with a quantity having only two values, spin up and spin down. In this way I can have ten thousand of these spins without overwhelming the computer. The interaction doesn't distinguish up before down, but the fact is that neighboring spins prefer to have the same direction. The final result will be that either they're all pointing up or all pointing down. Which alternative appears is sensitive dependent on the initial conditions. If, out of the ten thousand, I switch the initial values of two spins far away from each other, then one time it will result in one way and the next time in the other.

Heide: That explains the meaning of "coincidence" and "necessity" within the framework of this simple system. It's necessary that a star be built. These areas are so dominant in the state space that we will never be able to find an initial value that doesn't result in a star, even if such trajectories exist mathematically. When and where the star settles is purely random, because these trajectories can't be separated with enough calculative precision at the beginning.

Harald: Then it wouldn't really help for me to make my calculations at a rate of 10^{-17} instead of 10^{-16} exactness. This would only prolong the time period after which I can't say anything with precision by 17/16, and afterwards we'd be in the same situation. We're up against a wall that we'll never be able to climb over.

Me: After having explained the meaning of coincidence and necessity in this simple system, we should think about if we couldn't be able to find even embryonic forms of an ultimate goal.

Heide: How can you presume that something like that exists when you yourself said that the individual test particles are running around aimlessly? How could there be a plan behind the system as a whole?

Me: But that's what seems to be happening. What we've called the probability of a configuration—namely, the volume in the state space where this configuration is realized—is essentially the entropy of this state. It is the yardstick for the number of configurations. Events are globally dominated from the urge to create the configuration with the greatest entropy. That's why the larger stars swallow the smaller ones; entropy is something like Darwinist fitness. The Ω-point of Teilhard de Chardin is, in our model, the state with only one omnipotent star that everything else is moving toward.

Heide: What you've just said is simply an imaginative way of paraphrasing the simple statement that the system is moving toward equilibrium.

Me: Of course you're right. It's just that equilibrium here has an unusual structure. You'd think that the state of greatest entropy would be completely disorderly, that the particles are evenly distributed throughout the area. Here, we've got just the opposite: they're all concentrated in a star, but this is then balanced out by the fact that the star is so hot and therefore the speeds are all disorderly. But I can still interpret the course of events teleologically (meaning purposefully) as everything, necessary and random, is striving for a certain order so that the ultimate goal of this creation, a large star, will be formed. This final state is a necessity, while the individual fate of the stars is random. We've seen that the time of death of two stars or the moment when they merge together is dependent on their exact position. The error margin due to rounding off prevented us from predicting these events. In reality, there are no rounding-off errors, but there are always tiny unpredictable disturbances, which have the same effect. In our simple model everything is closely related: randomness and necessity, ultimate goal and natural selection.

Our computer studies have led us to problems that have long been subject to heated debate among philosophers. Naturally, no simple particular case can offer conclusive solutions. What it can do is show that some things have to be logically compatible, because they come from the same mathematical source. It's ridiculous, anyway, to say of two mathematically equivalent descriptions that one is correct and the other wrong. At most you can say that sometimes one perspective is called for and sometimes the other. That's the reason I find it strange that a marked hostility toward teleological observations can be found in some biologist circles, while some physicists only think in teleological terms any more. The reasons for this stretch back over centuries when it was discovered that the actual trajectory of a particle distinguishes itself from all possible trajectories in that the integral of its so-called action turns out to be the smallest. A particle always has an eye toward events in a random future. Feynman put this fact to great use in quantum mechanics, and it's now an essential part of modern theories. The best thing for me to do would be to summarize the facts of our model and withhold all comments about my preferences.

a) There is strict determinism. Each point in a high-dimension state space has exactly one trajectory that goes through it and that tells where the state will be for every time interval.

b) Human measurements can be made not of points in the state space but of a somewhat larger area. The trajectories generated agree after some time to build a star. Where it will be after a certain period of time is still open. a) and b) are the sources of necessity and randomness.

c) As after a longer period of time, there will always be a star lumbering around somewhere; this can be taken as the ultimate goal. Wherever the star is, it is in the state of the greatest entropy. It doesn't have any influence on where the star is exactly at that time; entropy is only dependent on the size of the star.

d) We always began with a starless state—that is, with a low level of entropy. On the way to the final state, structures with smaller entropy—that is, smaller stars—were always superseded by structures with greater entropy. If we were to replace the word "entropy" with the word "fitness," we would have an event like natural selection in biology. If we only look at one aspect, like the spatial distribution of the particles, it seems as if the order of the distribution increases and the entropy decreases. The most disorderly state is uniform distribution, and the most orderly is when all particles are united in a ball, which is actually what the system aims for. This agrees with the increase in the total entropy, because the temperature rises and the disorder of speed also increases. When, during the creation of life, the order of a part of the matter increases, it doesn't have to contradict the second rule of thermodynamics, the increase in total entropy.

THE SUN: SISTER OR GODDESS?

What makes the sun a sun god?

Our sun plays a special role in all mythologies; it is the source of all life, after all. The fact that deep on the ocean floor, where magma seeps out, single-celled organisms are able to thrive in intense heat, enormous pressure, with only sulfuric fumes as nourishment and they don't need the sun at all to live is irrelevant for peasant cultures.

It's the sun that ensures growth and survival, and it was raised to the status of a goddess. (In countries with a lot of sunshine, the sun is masculine and became a god.) In ancient Egypt, the sun god Ra held a high place in the hierarchy of the gods. Therefore it's a little surprising that St. Francis of Assisi praises the sun as a sister (or brother to be exact) in his famous "Song of Brother Sun." It is represented as something like us, subject to the same laws, holding the same powers. Doesn't this seem a tad disrespectful? After all, the sun has performed a miracle in the creation of life and is legitimately a goddess. We would like to explore this question in this section, and we come up with the clear answer of "somewhat." If we look at the situation from the point of view of macroscopic physics, the Egyptians were right. The sun has a property, namely, a negative specific heat, which is not allowed by the laws of thermodynamics that we have to obey. The sun, therefore, is breaking a law of nature, and thanks to this miracle, it can maintain sunshine and give us life. On the other hand, if we think of the sun as a mass of atoms attracted to each other by gravity, then we can conclude from the laws of motion that, when sufficiently large, the sun gains this ability. Then it's our sister again, albeit a very, very big sister. If we dig deeper and try to trace this result back to the elementary particle level, we once again come across several coincidences, and the whole thing becomes even more fantastic. Of course, we're still a miracle even at this level, and we can go ahead and agree with St. Francis.

Let's start by discussing the sun's extraordinary abilities. The specific heat is defined as the change in temperature with the energy. Positive specific heat just means that when I supply a body with energy, it gets warmer; if it loses energy, it gets cooler. Negative specific energy is just the opposite. This last fact should alarm the discerning reader. I will grant the right to object and open a question-and-answer period.

Discerning reader: A body with negative specific heat warms up when it loses energy. Then it would always be releasing more and more energy. This would solve all of our energy crises. I'll have to patent this!

Author: But if you come across an educated clerk at the patent office, you'll find your application rejected with the comment that Elliott Lieb and Joel Lebowitz had proven that normal matter without dominant gravitation always has positive specific heat.

Discerning reader: You said that this was always a condition for thermodynamics. Then it must be pretty obvious and easy to prove.

Author: That's not quite true. The proof requires all the finesse of quantum theory, and even that's not enough. If the electrons were bosons, then there would be negative specific heat.

Discerning reader: Still we've got the positive kind, and the whole thing seems like an imaginary bogeyman set up to scare off amateurs.

Author: Not at all. Cosmic bodies dominated by gravitation always have negative specific heat. As more than 99 percent of visible matter is found in stars, negative specific heat is the rule, and we are the exception.

Discerning reader: But doesn't a body with negative specific heat come to a bad end?

Author: When the body isn't too strongly dominated by gravitation, then it can improve and mend its wayward ways. But a star that's too big will never escape.

Discerning reader: I thought that in this case it would end as a black hole, and at least there it would have a break.

Author: Yes, that's what was thought until Stephen Hawking showed that the disease of negative specific heat in an advanced stage is incurable. Even a black hole shines and becomes hotter and shines some more until . . .

Discerning reader: Until?

Author: Until the end.

Discerning reader: What does that mean?

Author: Until it has reached the greatest possible temperature.

Discerning reader: And that is?

Author: The temperature corresponding to a Planck's mass. (According to Einstein, a mass corresponds to an amount of energy, which, in turn, according to Boltzmann, corresponds to a temperature.)

Discerning reader: And what happens next?

Author: A few more photons with Planck's energy slip out, and then there's nothing else.

Discerning reader: There has to be something left!

Author: Absolutely nothing.

Discerning reader: But a black hole is such a horrible cut in space; it has to leave a scar somewhere.

Author: I would have to know a bit more about quantum gravitation to be absolutely certain, but as far as I know, space heals and everything disappears.

Discerning reader: That would sure be a shame for our sun.

Author: I can offer a little comfort. Our sun will only make it to a third-class funeral, which doesn't call for a grand firework finale.

We have reentered the enchanted forest of physics and need to find our way back to reality. We actually already saw negative specific heat on screen previously, in discussing the birth of a star. If we start with a somewhat greater amount of energy, then the beginnings of a star vaporize and everything remains steady. With a lower energy amount, stars are built and the particles are extremely hot inside. Lower energy means higher temperature, and that is the phenomena of negative specific heat. To those who find that this deep question is being dismissed out of hand, I can offer simple theoretical evidence. The temperature T is linked to speed in such a way that it is essentially the kinetic energy E_{kin}. To be more precise, for N particles in the appropriate units $E_{kin} = (3N/2)T$. For motion in the gravitational field at time average E_{kin} is even overcompensated by the negative gravitational energy E_{grav}; $E_{grav} = -2E_{kin}$ ("virial theorem"). For the total energy $E = E_{kin} + E_{grav}$ it follows that

$$E = -E_{kin} = -(3N/2)T,$$

and we already have the negative specific heat: a deeper, more negative E pulls along a greater T. Now I've overshot the mark a little; as the discerning reader will ask, if that's so easy, why didn't we learn about it in school? It even took twenty-five years after the discovery of quantum mechanics before Lieb and Lebowitz were able to show that it doesn't hold true for normal matter after all. This is a fascinating chapter in mathematical physics, but it's so peppered with mathematical details that it exceeds my ability for breaking things down for a broader audience. In this case I can only offer a bit of biography.

First a little about Lieb and Lebowitz, who were the first to demonstrate thermodynamic stability—that is, the positivity of specific heats—for systems consisting of electrons and atomic nuclei. By demonstrate, I mean that they proved that the positivity is a result of quantum mechanics and the nature of the forces between these particles. In doing so they penetrated a deep secret of matter and deserve to be introduced.

Joel Lebowitz was born in Western Ukraine. This multicultural country was often the plaything of various nations, including the Poles, the Russians, and the Austrians. It was taken over by Hitler in World War II, who wanted to exterminate all of the Jews living there. Despite his young age, Joel was deported to Auschwitz, and you can still see the concentration camp number tattooed on his arm. But he survived, somehow.

As a little boy, he was able to slip through holes in the murder machine. He was liberated in 1945, and along with lots of other floating bits of world history, he washed up on the American shore. The question was how he was supposed to earn his living, as he had hardly received an adequate education in the concentration camp. With the intelligence and perseverance that let him survive Auschwitz, he was able to work his way up in academia. At Rutgers University he ultimately led one of the most internationally renowned research groups on statistical mechanics. His scientific success did not mean he forgot the horrors of his youth, and he also spent a great portion of his time and energy on supporting innocent victims of political persecution. He recognizes no ideological boundaries and has no regard for his personal safety, as evidenced by his attendance at dissident gatherings in the Soviet Union; he doesn't care about his colleagues' reluctance when he introduces this topic at every international conference. He just sees how his own fate is still being suffered by many even today.

Elliott Lieb and I share not only a long professional collaboration and friendship, but our names are also bound together in a mathematical theorem: the Lieb-Thirring inequality. First, let me clear up a possible misunderstanding; an inequality is not a false equation, but rather states that one of the sides will always be greater than the other. Perhaps I should relate the story of how we discovered it, as this shows scientists as hunters: the suspected result hides somewhere else like a shy deer each time. You've got to stalk the prey for a long time, in our case for many years, until your time has come and you can capture it.

I met Elliott in 1968 at a conference and was immediately impressed by his style. He presented his results concerning the rest entropy of ice, which shows how many molecular instructions are needed for an ice crystal to be specified. This problem had occupied geniuses for decades, but remained unsolved. The double Nobel Prize–winner Linus Pauling had published his speculations, but couldn't prove them. Elliott saw this as a mathematical problem and took the attitude that you could either prove a result, or you shouldn't bother. He was successful in the proving, which, by the way, did agree with Pauling's speculations. They, however, dealt with the problem in just one plane. The problem of the three-dimensional tetrahedron grid used as scaffolding for ice is still waiting on a Super-Elliott for its exact solution.

Figure 4.4: Elliott Lieb and myself at work.

Our first meeting was brief and bore no scientific fruit. At that time I was one of the directors at CERN, the European lab for particle physics. The next year I learned of the work by Lieb and Lebowitz mentioned above, about the thermal behavior of normal matter. This seemed to me to be so essential for general knowledge in physics that I invited Elliot to give a lecture at CERN. When I picked him up at the airport and saw that he had only a small zippered case with him as luggage, I asked where he had his notes for the lecture. He answered by pointing to his forehead. His brilliant lecture given off the cuff impressed almost everyone. Personally, I wasn't entirely satisfied because his proof of thermal stability was based on another kind of stability. It says that the energy to be gained from a conglomerate of atoms when new atoms are added cannot increase more strongly than the number of atoms. For example, two liters of gas contain just twice as much energy as one liter. This kind of stability had been proven a few years earlier with some spectacular mathematical fireworks by Dyson and Lenard. They used forty pages of highly refined inequalities and tricks, but it wasn't easy to comprehend. I confessed to Elliott that I still couldn't really keep track of it all, and he agreed. He also wanted to have a more transparent proof of this stability, but he just didn't know how. We added this question to our "things to think about" lists.

Although we were able to solve related problems in the next few years, we didn't get any further on this one. I returned to Vienna. In 1974, when I was able to allot a guest professorship, I invited Elliott in the hopes that we would find a simpler proof for the stability of matter. His stay was very stimulating and productive, but the desired result remained elusive. We stayed in the blind for another year until Elliott could return to Vienna in 1975. When I picked him up at the airport he said, "I feel it in my bones; this time we will lick the problem." This was followed by weeks of intense work, chalkboards filled with innumerable attempts, which were all then erased as being of no use. Then just a few days before Elliott was to return home, it suddenly stood before us; the Lieb-Thirring inequality. It states that the Pauli's exclusion principle forces the kinetic energy of electrons with their density to increase to a power of 5/3. From there, with the results that Elliott Lieb and Barry Simon had already proven, you even get the stability of matter.

Our proof was not only a significant simplification of Dyson and Lenard (we only needed three compared to forty pages), it was also more precise numerically. With so many estimates and inequalities, something is always lost a little, about a factor of two per page. With forty pages this adds up to 2^{40}, that's around 10^{14}. Dyson and Lenard could only show that, in a pile of N atoms, there can be no more than $N \times 10^{14}$ times the energy of an atom. We were able to reduce this insane factor of 10^{14} to 8, but that still wasn't optimal. The empirical evidence said that the factor couldn't be much larger than 1. That's what we got when our inequality would hold in its most extreme form. Our inequality says that the kinetic energy of electrons must be greater than a certain factor times the integral over (density)$^{5/3}$. The strongest correct statement was made when the factor before (density)$^{5/3}$ was $(4\pi)^{2/3}$ greater than what we had. So we introduced this stricter, unproven form of the inequality as a conjecture. While our inequality was hauled off to the museum, our conjecture became a favorite plaything among experts. Many have attempted with their genius to prove it correct, others to refute it with supercomputers, but neither has worked. It was only possible to gain a few percentages for the missing factor. One day, after decades had passed, Timo Weidl, a man I hadn't heard of before, found that the problem just needs to be taken from a slightly different angle and then the missing factor $(4\pi)^{2/3}$ melts down to $2^{2/3}$. It is written in the stars if this can be further reduced to 1 and turn the Lieb-Thirring conjecture into a theorem.

Of course, the Lieb-Thirring inequality holds true for stars as well, but its conclusion about stability fails with gravitational forces. This opens the door to negative specific heat, which already made an appearance at the beginning of the twentieth century in a book by Robert Emden about balls of gas. He didn't trust himself to emphasize this paradoxical behavior, and it's more or less suggested when reading between the lines. It also appears as a thrown-in remark in the famous textbook series by Lev D. Landau and Eugenij M. Lifschitz, but without being discussed further. This idea was never able to catch on in physics, while it became common knowledge among astronomers. In 1968, Donald Lynden-Bell and R. Wood dedicated a detailed study on exactly these questions, for which Lynden-Bell was even decorated with an important British award.

I was confronted by the problem when, during a lecture by Fred Hoyle, I was startled by the comment, "A star is like a heat machine: the more energy it radiates, the hotter it gets." This meant exactly negative specific heat. I had always proven in my lectures that the specific heat could be given by the fluctuation square of energy. As a square of a real quantity, it must be positive. What was going on here? After much consideration, I realized that my proof only referred to a system in a heat bath. A system of negative specific heat, on the other hand, would never offer itself for thermal balance with another body. This quirk can only be cultivated in isolation. This does not, however, prove that an isolated body is able to have negative specific heat. I tinkered with different mathematical models until I came to a simple realization. I played dumb and told myself that gravitational interaction is something like a hole, only that every particle attracts every other one. The depth of the hole should therefore be proportional to the number of particles N. I ignored the interaction among the particles and placed a hole with a depth of N somewhere. I could explicitly calculate the thermal properties of this primitive model and look; for a certain level of energy in this model, there was negative specific heat.

I was thrilled. I copied everything down and handed it in to be published. My enthusiasm did not infect the referee for the magazine, and he didn't want to accept the study. Not that he said that it was too trivial, as the astronomers knew this already. On the contrary, he thought that it was too revolutionary and that there must be something wrong with it. He was a gentleman and didn't just reject it outright, but invited me to give a lecture. He invited all of the experts in statistical physics, who then pressed me to abandon my silly ideas. I stood there like Martin Luther in front of Emperor Charles V and could only say that I cannot recant anything against my conscience and knowledge. I didn't have to say "God help me, amen," and I didn't have to flee to the Wartburg. Still, the referee was intent on quelling this heresy. When he couldn't find any mistakes in my math, he said, "If what you say is true, then the conditions which flow into statistical mechanics aren't met. The problem is one of dynamics, and you aren't able to show that your thermal properties have anything to do with the actual motion of the particles. I find the whole thing to be nonsense, but if you want to make a fool of yourself, then publish it." I took the risk and published.

In the following years my collaborators and I were able to make the model more true to life. The negative specific heat remained, but so did the objection of the referee. After all, who would be able to calculate the motion of so many particles attracting each other? Twenty years later the tide had turned and the computer began its era of triumph. The calculation of the motions for four hundred particles all attracting each other was no longer a problem, and three Dutchmen—A. Compagner, C. Bruin, and A. Roelse—studied this case (our undertaking described earlier in this chapter was just a continuation of their work). They found confirmation of my results: a star was born, and the lower the total energy, the faster the particles became and therefore the hotter it got. Had the referee made a fool of himself? Not entirely, and besides, he wasn't only a gentleman, he was also a Nobel Prize winner. The computer showed something that had to escape my thermodynamic observation. As described previously, it first gets much hotter in a star than it does outside, and it's only much later that the temperature evens out. This first condition is our source of life, as we could not evolve when the temperatures are equal. Heat death was not only initially avoided, but the exact opposite situation developed. The initial apparent balance was not stable, and there came to be hot spots. With positive specific heat, the stars would not have warmed up after radiating away their gravitational energy, and in their cores the nuclear reactions would not have ignited. Then there would be a world without C, N, and O, and it would be lifeless.

The sun was returned to its rightful place: as a sister, because everything happens according to the same laws of nature that we're subject to. But also as a goddess, the source of our sunshine and also of life.

THREE TYPES OF FUNERALS

Some stars die quietly, while some go out with a bang.

When cosmic gas masses come together in a ball, this creates a sirocco of unimaginable heat. We saw a direct demonstration of this blazing heat when the Shoemaker-Levy comet crashed into Jupiter. Hours later, the tiny crash sites were still visible as blinding spots on the face of Jupiter. This kind of surface heat gets used up quickly, and the show was soon over. In a star, the heat is sitting inside and it takes an unbelievably long time until the radiation can get out of there. If a photon were to fly directly out of the sun, it would only take a few seconds to get from the center to the surface, as the sun's radius is 10^9 m and the speed of light 3×10^8 m/s. In fact, this takes millions of years; we're enjoying the sunshine that was created eons ago.

How did we figure that out?

It's quite easy; we've already done it. The problem is analogous to the marble game in appendix 6. We've just got to replace the words "black" and "white" with "forward dispersion" and "backward dispersion." The photon in motion is constantly being scattered, and for the sake of argument, let's say that it's equally likely to go forward as backward. If it is scattered N times, as we discovered in appendix 6, then the probability, that the photon goes forward $N(1 + d)/2$ times and otherwise backwards is 2^{-cNd^2}. In order for this probability to be discernable, we need $d < 1/\sqrt{N}$, with N representing the number of dispersions. For $d = 1/\sqrt{N}$ the photon moves $N/2 + \sqrt{N}/2$ times forward and $N/2 - \sqrt{N}/2$ times backward; it therefore progresses by $\sqrt{N}$ altogether. When there is about one dispersion per mm, then the photon will only progress by $\sqrt{N}$ mm instead of N mm after N dispersions. For N this means that $\sqrt{N}$ = sun's radius in mm = 10^{12}.

This huge figure for the sun's radius can be estimated in the following way: the sun is approximately 100 times as big around as the earth, which has a radius of around 10,000 km, so that the radius of the sun would be 1 million km. A kilometer has 10^6 mm, and when everything is written in decimal powers, then $10^{6 + 6}$ mm = 10^{12} mm. The escape route takes $N = (10^{12})^2 = 10^{24}$ dispersions. For one mm it takes a photon ~ 10^{-11} s, so for 10^{24} dispersions 10^{13} s, which is several million years.

It's no accident that this period of time is the previous Kelvin estimate of the sun's age. If the heat of the sun were not fed by nuclear reactions, it would have cooled down after a few million years. How can nuclear reactions take place when the nuclei are dressed in electron clouds and, in normal matter, never meet? It's like this: when the gravitation becomes too powerful, the atoms are squashed, and this makes a mush of electrons and nuclei no longer covered by their own electrons. This kind of mush is called "plasma," and that's where the nuclear reactions listed at the beginning can take place. Through gravitational pressure, hydrogen in the sun's core has the density of 150, more than 10 times denser than iron, and the atomic nuclei are much closer together than normal.

It depends on the star's mass if the gravitation is able to compress the atoms into plasma. At masses greater than Jupiter's, the brittle atoms begin to shatter under the enormous pressure. For a nuclear reaction, the Gamow factor we mentioned earlier still needs to be overcome. For this to appear at a respectable rate, we need greater densities and a much greater mass than Jupiter's. The reactions only occur deep inside the sun where we can't see them. How can we be sure that they exist? Today, we can be sure, because we humans have gained a new sense, thanks to neutrino eyes. These are enormous detectors that can see neutrinos and in fact detected the νs of the reaction $P + P \rightarrow D + e^+ + \nu$ inside the sun. The νs are so solitary that they are able to slip out of the sun unhindered, which then only takes a few seconds. They fly to us fresh, without having their energy cut to pieces like the photons. In Japan, there's a neutrino eye that's direction sensitive, and it could even show that the neutrinos in fact come from the sun. This is an excellent validation of Atkinson's and Houtermans' vision, and now it's time for me to introduce the latter.

Fritz Houtermans' fate was sealed by the political insanity of the twentieth century. After his schooling in Vienna, he worked in Germany. He left when Hitler came to power, partly out of conviction and partly due to the fact that he had some Jewish ancestors; his family tree wasn't "pure." At that time there was a liberal period in the Soviet Union, and he got a job in Kharkov. This period was not to last long, and the Stalinist terror began. Houtermans was deported to a gulag where he was subjected to

Figure 4.5: Fritz Houtermans in Palermo, 1961

torture and interrogation. He was released only after the Hitler-Stalin pact, but only out of the frying pan into the fire. In Germany, he was stuck in the Gestapo prison as a suspected communist, and it's thanks to the efforts of Max von Laue that he came out alive. Physically and psychologically shaken, he would try to calm his nerves with alcohol, coffee, and nicotine, which only ended up destroying his health. Despite all this, he constantly was coming up with original ideas about all kinds of things. I once asked him how it was for him after the war in Göttingen when there wasn't anything available to smoke. He replied that he bought himself a pipe and wandered around with it at night looking for English soldiers who were smoking cigarettes. He would ask them for a light and held out his empty pipe. When they put their cigarettes in the bowl, he would inhale deeply several times and in this way get a few drags on a cigarette. One time he said to me, "Einstein's statement that an excited atom will be more likely to release a photon, the more of these kinds of photons are flying around, means that there must be a kind of light avalanche when there are a lot of excited atoms together. The more photons that are already there, the more the excited atoms will release the same kind, so that the light intensity increases exponentially." This was long before the invention of the laser, but I have never again heard such an apt description of how a laser works. Neither he nor I knew exactly how this could be achieved, and the laser was developed years later in the United States.

Over the years our relationship developed from master/servant to friendship. We even became something like a team. We published a joint study together; it wasn't earth-shattering, just earth-quaking; the earth was still whole afterwards. Houtermans was always interested in finding out how he could prove his idea that solar energy = nuclear energy. Once he came to me and said, "You're so good at figures. Don't you want to see if, out of the hailstorm of solar neutrinos, it wouldn't be possible to show up a few?" It had to be a dense hail; this I knew as Fermi once had estimated 10^{11} neutrinos per cm^2 and second. How could it be possible for us to catch them when even our huge sun couldn't? I came up with various mechanisms, but didn't find anything useful. At that time the neutrino was purely a thought construct invented by Pauli. It hadn't been experimentally proven, and many of its reactions were unknown. As usual, Houtermans and his visions were ahead of their time. At least we were able to discover that neutrinos with a high energy had to appear much larger, but that didn't help; accelerators that could have produced them hadn't been introduced yet. Today they do exist, and they are also able to produce many particles with high energy at the same time. Neutrinos have become an important instrument of high-energy physics. There are even plans to produce a neutrino beam in Switzerland and shoot it straight through the earth to Italy to experiment with there. At that time this would have seemed like science fiction to us. Unfortunately, Fritz Houtermans wasn't able to experience this progress. His health was ruined, and he only lived to reach a little over sixty.

I don't know if any of the Green parties have distanced themselves from solar energy, as, in the end, it's nuclear energy, too. You could say that using it to heat up the sun would finally be an environmentally friendly way to use nuclear energy. I would have to contradict you, not due to ideological reasons, but from the scientific point of view. Nuclear energy is not heating, but cooling the sun; the negative specific heat is once again standing everything on its head. Without nuclear energy, the sun would compress even more, and in doing so, become even hotter. But because the nuclear reactions in the sun's center feed in as much energy as is radiated away on the surface, the temperature remains constant. It would only increase if the total energy were to decrease. This will only happen when the hydrogen in the sun's center is burned into helium. There will also be further nuclear meltdowns that produce carbon, oxygen, and increasingly heavy elements. They have a larger Gamow factor and need greater temperatures. The sun will therefore have no choice but to tighten its belt and use the freed energy to ignite the next level of nuclear reactions. These will last it for a few more billion years and then comes the next period of fasting, and after the successful diet the next elements will be cooked up. At some point this also has to come to an end, and this happens with iron not at the end of the periodic table of elements. After iron, there can be no more energy gained in further nuclear meltdowns, and the star has no choice but to declare energy bankruptcy. What happens next is determined by the cosmic bankruptcy court, which is merciless. It knows only the death penalty. It does allow for three kinds of funerals, depending on the weight of the delinquent.

Third-Class Funeral

The most modest ceremony is granted stars which are only about as heavy, but no heavier, than the sun. Due to their lavish energy lifestyle, they've already had to go through several starvation diets and end up as small as the earth, as so-called white dwarves. The gravitational pressure at the core is not yet too overwhelming, and the plasma from which they're made can resist it. They pass away peacefully in their sleep, radiating the rest of their energy and become brown dwarves before they completely cool off. If they're light enough, they end up frozen as crystal. This means that the electrons build a homogenous charge background where the nuclei are organized into a normed grid, a so-called Wigner crystal. The diamonds found on earth, which are, after all, also created under enormous pressure, will grow to the size of a finger at most. Cosmic nourishment packs a more powerful punch; it can stamp an entire star into a crystal the size of the earth, but with a million times more mass. Even a finger ring made out of this kind of crystal would weigh tons. All in all, becoming a crystal like this would be a dignified way for our dear sun to go.

The name on the grid stands for a universal thinker, Eugene Wigner. I'd just like to mention one of his quirks in passing, his opaque sense of etiquette. At one seminar he noticed a mistake on the chalkboard; there had been written + instead of –. Wigner didn't want to make the speaker feel too uneasy, so instead of saying, "The sign is wrong," he just said, "It seems to me that this sign is not very good."

Second-Class Funeral

A bit more of a pompous ceremony is granted to stars that are a little, but not too much, heavier than the sun. They have had a more wasteful lifestyle and had shrunk even more inside. The distance between their electrons went from 10^{-10} m, not to 10^{-12} m as with the third-class stars, but compressed down to 10^{-13} m. This is exactly their Compton wavelength, and as we learned earlier, due to quantum mechanics, electrons begin to whirl around almost at the speed of light in this predicament. This brings about an effect for which Einstein, or rather, his special theory of relativity, is responsible. The electron framework becomes softer and its energy contribution increases with its density ρ not as $\rho^{5/3}$, but from now on only as $\rho^{4/3}$. This is the same power of ρ with which the gravitational energy turns negative. As the positive and the negative source are equally strong, there is no longer a state of equilibrium; one of them wins—which one depends on the size of their preceding factors. This is always the gravitation with sufficiently large stars, and the more ρ is increased, shrinking the star, the more energy this generates. The total energy of the star decreases during this process.

The exciting story of this dramatic instability can be quickly relayed. It first appeared in a study by I. Frenkel about metal electrons. It was published shortly after the discovery of quantum mechanics. No one had suspected it in that context, and it hardly registered, but was often being rediscovered—for example, by R. Fowler, among others, apparently independently of one another, as none of these authors referenced their predecessors. Five years later, Lev D. Landau and Subrahmanyan Chandrasekhar derived the same result once again. This was unsettling to Landau, as he realized that, according to this criterion, many stars would have to be unstable, but they were still shining. He backtracked and borrowed the occult idea of Niels Bohr that the energy is not conserved in some reactions, and for that reason the energy argument didn't seem conclusive. Chandrasekhar stood by his result and was often ridiculed because of it, even by such prominent people such as Eddington. He later received the

Nobel Prize, but that would take some time yet. At first it had just been demonstrated that energy can be won by shrinking stars smaller and smaller. This raises the question: what should be done with all this energy if it takes millions of years to squeeze a photon out of a star?

Gamow realized that nuclear forces would have to play a role, not as an energy donor, but as an energy user. That's just what we need as we're trying to get rid of energy. Along with Mario Schoenberg, he thought of a process which would do just that, and they named it after the Brazilian casino Urca, where money also disappeared in no time flat. The simplest way is the reaction $P + e^- \rightarrow N + v$, where it's necessary to pay the difference in mass $N - P$. There's a neutrino in change, which can leave the star in a split second. If the energy of the electrons becomes greater than this difference in mass, they can crawl into the protons, and a neutrino guarantees that the deal is energetically correct. Only neutrons are left over, and the electron frame that held up the star against its gravitation is gone. The star collapses on top of itself, the neutrons fly in a free fall toward the core, and due to the enormous gravity, the collapse takes place in a split second. This leads to the cosmic catastrophe known as a supernova, where a star lights up with such brilliance that it outshines its entire galaxy. The neutrons are either thrown back by the neutrinos flooding in the other direction as they're speeding toward the core, or they fall into the neutron lakes created in the core and splash back. The energy per particle is as great as in a nuclear explosion. And we're not talking about just a handful of matter, but an entire star, which is 10^{31} times the size. An unfathomable heat is generated, which is the source of the illumination. When all of the excitement has died down, there's a neutron star left over, a star made completely of neutrons, but with the density of nuclear matter. It's a ball with a radius of about 30 km. Due to the omnipotence of gravity, its surface is polished smooth; it might have a few elevations, but they're less than 1 mm high.

You may ask yourself how anyone can know all this, when this infernal event is covered by burning gas clouds. The neutron star signals all this to us when it becomes a pulsar. It sends out bundles of light, and these show us how often it's turning on its axis, similar to how a lighthouse works. It's not easygoing like our sun. These cosmic pirouettes occur 30 times a second, with the star having become incredibly small. This means that it can't even be as large as the earth, as otherwise its circumferential speed would be faster than the speed of light, and it would have long been torn apart by centrifugal force. How is it possible to know that the neutrinos are actually able to escape? Our neutrino eyes should be able to see this when they're not sleeping. In fact, a supernova flared up a few years ago and while a neutrino eye was open. At the same time as the supernova was seen to flash, our neutrino eye registered seven neutrinos within seconds, a very unusual event. Gamow really did seem to have identified the cosmic plan.

A valid question would be to ask who actually profits from these cosmic executions. Did the Lord in his wrath want to dethrone and humiliate the arrogant? No, he wanted to raise us, humanity, to life, and for this, no sacrifice was too great. He is powerful, or even violent? It's not our place to quarrel with him; in any case, without supernovas we wouldn't exist. The heavy elements essential for all life—essentially just waste products of cosmic nuclear reactions—couldn't be created in the rush of the big bang. They had to roast over the coals of the sun for billions of years before they were done. If there were only third-class funerals, then they would be forever locked inside and no miner would be able to chip them out of the star crystal. During a supernova, the star explodes, just as the fruits of some trees only break open during a forest fire to sow their seeds. In this way a piece of star slag is roasted so well that all elements lighter than iron are basted out. In the last few seconds they're also irradiated with neutrons so that the palette of chemical elements is completed before the gigantic explosion propels them into the universe. Millions—or was it billions—of years later, they entered a quiet area, gave birth to our planets, and blossomed into our earth.

First-Class Funeral

The most radical funeral has been arranged for stars that are a bit heavier and have more than ten times the mass of the sun. These stars have the most wasteful lifestyle; they shine the most daringly in the sky—like Sirius. Woe to you blasphemers, the God of physics is even mightier than that of the Old Testament. You should get out of his sight, as he is not only able to create something from nothing, he can also dissolve something into nothing!

Naturally, stars this heavy aren't spared the supernova fate, but the neutron star this leads to is too large and therefore unstable. Not even the nuclear matter can stand up to the gravitation, and everything is sucked into the insatiable maw of a black hole. We don't even need to worry about who will take the energy away. Everything disappears into this cosmic trash can. It then becomes the burial plot, the eternal peace of the Schwarzschild space. Finally, the Hawking radiation leaks out. It comes in drips over billions of years, but it slowly consumes the mass of the black hole until there's a grand finale of fireworks that extinguishes all.

What was the meaning of all this? Some physicists have continued with their fiction and say that the more productive a universe is, the more black holes it creates. I'll get back to this idea, but would like to stop for now; we've been indulging in enough romanticism.

THE MORAL OF THE STORY

The life of a star is based on different, yet equally ingenious, principles from ours.

Stars, from Ra the sun god to the smallest spark in the blue night sky, have shown themselves to be highly complex beings. They have an interesting birth, a long life, and each one has its own fate and its own death. They live according to three forces: the electroweak, which causes electric and magnetic phenomena and β-decay; the strong forces that connect particles and atomic nuclei; and gravitation. For stars to flourish, these forces must be coordinated to the highest degree. The electrostatic force can't be too strong; otherwise the hydrogen nuclei would repel each other too strongly and couldn't melt together. It can't be too weak, either, or else the atoms would be crushed by masses too small, and the stars wouldn't reach the right heat we need for roasting elements in our alchemists' lab. Sometimes the highest precision is demanded, as with the $^8Be = \alpha + \alpha$ (we're now once again using the common abbreviations: α = atomic nucleus of helium, Be = atomic nucleus of beryllium, C = atomic nucleus of carbon). As we've said, two helium nuclei will not stick together forever, but at a certain amount of energy, for a little while. A third has to join in as soon as possible to come up with $^{12}C = \alpha + \alpha + \alpha$, but this must be done carefully without the electrostatic repulsion being too large between the α, or else it won't work. By putting up parameters, you see that there's only a small peephole in the parameter space through which the heavier elements can slip into the Promised Land. Everything has been expertly prepared for us to prosper. The ^{8}Be-hole is there so that the heavy elements' ascent doesn't happen too fast, so that there's time for everything to roast thoroughly and cook up enough ^{12}C. It's also essential that the same thing not happen again with ^{12}C and a sticking together in such a way that the ascent to ^{16}O progresses too quickly. Too much oxygen and life can't originate. At first it's just a waste product; it's only later that we become addicted. But now I've caught myself reading the universe's blueprint like a gourmet would a recipe and am rating it according to my whim.

For the Urca process to ignite, the energy of the electrons, approaching the speed of light, needs to correspond to the N-P difference in mass; the weak interaction that gives us $e^- + P \rightarrow N + v$ can't be too weak so that it goes quickly, but also not too strong so that the v escapes the star. Beyond that, it has to be strong enough so that the neutrinos flooding out are able to grab the falling neutrons, turning the implosion into an explosion. The developmental period of the stars, a few billion years, must be coordinated with the time it takes for the cosmos to develop, as well as the time it takes for biological evolution. This is the only way that the cosmic incubator can spawn life. The wish list is endless, filling volumes, yet every wish has come true. All of these enormous efforts put into planning, just so that you and I, tiny creatures, are able to live on a cosmic speck of dust called earth? It seems that nothing is too good for the crowning glory of God's creation.

5

What Newton Suspected about Our Solar System

Our Planetary Family

Even the tiny crumbs circling the sun, the planets,
have had fortune smile upon them.

THE FAMILY HISTORY of our solar system, of mother sun and her nine chicks, stretches back a long time, 4.5 billion years. It has been amazingly peaceful. Our understanding of why this has functioned like clockwork has led us to many an adventure. Our sun, the most prominent phenomenon in the sky, was always worshiped like a god, but the planets were just points of light in the night sky like millions of other stars. It's a credit to the observational skills of ancient peoples that they were able to discern that planets are different. In one night it's initially possible to see that the entire spectrum of stars turns rigidly halfway on an axis going through the polar star. Over a longer period of time, it's possible to discern how the planets move around. Venus is the most obvious, sometimes appearing as the evening star, sometimes as the morning star. This gave rise to the worry that the danger of colliding with another star must be great while moving around so much in the sea of fixed stars; nothing was known at that time about the distances involved. It seemed that there must have been a heavenly traffic code. In the Middle Ages, none other than angels was deemed suitable to enforce it; it was assumed that angels led the planets around their orbits. Even Newton, who discovered the laws of planetary motion, thought that they wouldn't be enough to provide order over time. God was needed to set things right occasionally. However, the mathematical scenario Newton developed had no place for setting things anew, and as Napoleon asked the mathematician Pierre Simon Marquis de Laplace where the dear Lord was in his system, he famously answered, "I have no need of that hypothesis." It's still not that easy to prove that a catastrophe won't take place, that two planets won't collide, or that one won't be thrown into the sun or even be catapulted out of the solar system. That's why there has constantly been proof of the stability of the solar system and then evidence that the proof wasn't completely airtight. Finally, it got to be too much for King Oscar II of Sweden, who was scientifically inclined. He offered a significant reward for anyone who could provide conclusive proof of the stability of the solar system. This attracted the best mathematicians of the nineteenth century, and numerous successes were announced. The jury finally awarded the famous French scholar Henry Poincaré for his work. His Swedish colleague Magnus Gösta Mittag-Leffler had just founded the journal *Acta Mathematica*, still renowned today, and wanted to have this very famous work from Poincaré grace the first issue. Everything seemed to be going well until Poincaré discovered a mistake in his thought process; his proof wasn't conclusive; it wasn't even proof. He immediately telegraphed Mittag-Leffler to withdraw the article, but everything had already been printed. Poincaré had to arrange for the first issue of *Acta Mathematica* to be destroyed at his own cost, which was so great that it turned out to be more than the original reward. In the end Poincaré did come out a winner; he included his new insights in a book on new methods of mechanics he published. His wasn't just a method, but influenced the very way we think about these things.

The proof that eluded Poincaré was found a few years later by the Swedish mathematician Karl Sundman for the three-body problem; the sun + 2 planets. He was able to show that the Newtonian equations set the movements for all times for almost all initial conditions. This does not mean that it's certain that it will never come to a catastrophe. It does not follow from Sundman's proof that two bodies will not collide. He was able to find only one other variable besides time through which the movement during a collision can be tracked mathematically. He wasn't able to exclude the possibility that one of the two planets would be catapulted out of the solar system; this could even be the typical turn of events.

The question of the stability of the solar system never ceased to be asked until it was finally shown to be not only unanswerable but also irrelevant. Jacques Laskar was able to set the previously mentioned Liapunov time—after which events are so sensitive dependent on the initial conditions that they are unpredictable—at ten million years. What happens afterwards is so sensitively dependent on the origins that further assertions are worthless.

Johannes Kepler could have already set a period like this using the following argument: when an orbit has a 10^{-7} different distance to the sun from ours, then, according to his third law, the time needed to complete a rotation around the sun is also 10^{-7} different from ours. After 10^7 years, this orbit will have drifted over to the opposite side of the sun from us. In this time, the uncertainty of its initial position has grown from $10^{-7} \times$ distance to the sun $=15$ km to 3×10^8 km, the diameter of the earth's orbit. But the shape of the earth is about 15 km away from a round object. A more precise specification of the initial position doesn't make sense in the Newtonian equations, as the more the precision increases, the more fine effects affect the motion. The dependability of predictions made by Newtonian equations for time periods longer than 10^7 years is therefore questionable. This has only become a problem since it became known that the solar system is more than 10^9 years old.

There are special orbits that will see no harm for all eternity, but a catastrophe could take place very close to them. The question needs to be rephrased: are there configurations of planets where we can be sure that they will stay out of danger for at least 4.5×10^9 years even with small changes in the initial values? It appears that such a thing exists. When people ask me why the earth has retained about the same distance to the sun for billions of years and in this way provided the climatic stability necessary to develop life, then I have to go back to the beginning with my answer: the earth has apparently had a good guardian angel who led our solar system to a stable region. These configurations are lonely islands in the sea of states. It seems highly unlikely that other solar systems could have enjoyed a friendly climate for billions of years.

The Kepler Orbits

*The calculation of the planetary orbits by Newton
was the first triumph in theoretical physics.*

"It really is moving," Galileo is purported to have said after having been found guilty of heresy. His discoveries laid the foundation for modern physics and ushered in the age of science. It took almost four hundred years for his case to be retried and for him to be rehabilitated. The revolutionary statement was actually empty in content; Galileo was damned by religious prejudice. It's only today that we are able to pinpoint the "moving," and in an unexpected way at that.

First of all, we are only able to recognize relative motions—that is, when distances change. If everything moves in the same way, then nothing changes and we don't notice the movement. Galileo had realized that a body remains in a state of steady motion when no forces act on it. The forces among its parts are the same whether the body is in motion or at rest. This is why we feel just as comfortable in a moving train as we do when it's standing at the station, as long as the tracks allow a steady speed without shaking. The fact that we are unable to feel absolute motion is the pièce de résistance of Einstein's special theory of relativity. Einstein's opponents have turned the word "relativity" into "lack of conviction," and yet his theory only serves to work out the absolute. The absolute is a favorite phrase of philosophers, but it's only here that it gets a well-defined meaning: the absolute is that for which it is irrelevant if it is observed from a frame of reference at rest or in steady motion.

Steady motion does not have an absolute meaning, but rotation does. Everyone has had the experience that it feels different to sit on a chair that spins than on one that doesn't; you can feel the centrifugal force. This doesn't mean that we can't describe our experiences from a rotating system, just that then we've got to allow for forces of inertia like the centrifugal force. They are often so weak that it's acceptable to ignore them. For everyday living, we can act as if the earth is not rotating. We're allowed to use a motionless earth frame of reference and say, "The sun is rising." We don't have to pedantically say, "The earth has turned another little bit on its axis so that we can now see the sun." This doesn't mean that the earth's rotation can't be measured; on the contrary, it's demonstrated by Foucault's pendulum. How this works can be most clearly demonstrated by installing a perfectly hung pendulum exactly at the North Pole (which Foucault couldn't). If we look at the motion from a frame of reference in which the fixed stars are at rest, meaning the earth is rotating, then the pendulum continues to swing in the same direction. Looking out from the earth, the swinging plane will have rotated exactly 360° in twenty-four hours. Of course, this motion can also be described in an immobile earth system, but then an inertial force needs to be included (in this case, it's the Coriolis force). It produces the rotation of the swinging plane. If Galileo had said, "It really is rotating!" then he would have been able to prove, in principle, his assertion to the Inquisition. Movement itself doesn't make sense; it's also not true that the sun is standing still. Our Milky Way is a spiral cloud, and we're stuck in a spiral arm that has already rotated ten times around the center. It's even subject to debate if our Milky Way is moving or not. The expansion of the universe means that galaxies are moving away from one another, but everyone thinks that it's the others who are moving away.

Absolute motion could be defined as long as we had something stable in space, something that would give us something to go by, something we could measure the speed against. At first glance, it doesn't seem that something like this exists. But this is where the unexpected comes in. The cosmic background radiation allows us to define an absolute motion. It's nothing concrete, more like quicksand, but with the help of the Doppler effect, we can measure if we are moving in relation to background radiation. When we move toward it, its frequency appears to increase; when we move away, then it decreases. In this way, we can directly demonstrate our motion compared to a cosmic standard. Because the earth is rotating on its own axis as well as orbiting the sun, our speed on the earth's surface is constantly changing in relation to the background radiation. We're talking about speeds of $10^{-4} \times$ the speed of light. Today this can be measured; the precision of frequency measurements of background radiation is at 10^{-5} of the frequency. With modern technology we might be able to reach an agreement with the Inquisition about the absolute meaning of "It really is moving" and objectify Galileo's offence.

Who isn't familiar with the nice legend about young Newton, who had a vision that has since come to dominate how we understand the dynamics of the universe—namely, he recognized the universality of gravitation.

It has also played a decisive role in our observations up to now. Newton was supposedly relaxing under an apple tree looking at the moon. Suddenly an apple fell and hit him on the head, which sparked the thought process that the force keeping the moon in its orbit around the earth must be the same as the force that dropped the apple on his head. We don't know if this story's true, but if it was made up, then whoever made it up did a good job. Newton published his ideas only at the age of forty-four in his *Philosophiae Naturalis Principia Mathematica* in 1687, which contained the basic laws of motion of points of mass. Stars were also thought of as points. Newton's laws have reigned in the mechanics of the heavens ever since. Only hundreds of years later were they somewhat fine-tuned by Einstein. To understand the workings of our solar system, we need Newton's second law,

Force = mass × acceleration,

abbreviated as:

$$(F = m \times a) \qquad \text{(I)}.$$

The next law specifies gravity between two bodies:

Gravitational force = G × mass of the first body
× mass of the second body/(distance)^2,

abbreviated as:

$$F = G \times m_1 \times m_2/R^2 \qquad \text{(II)}.$$

G in (II) stands for the gravitational constant we've been using and not known a priori. Newton's vision was that G is the same in heaven and on earth; the moon and an apple have the same G. To test this, we need the masses of the earth, the moon, and the apple, and the distances earth-apple and earth-moon. I can't quite figure out how Newton could have known all this given that he didn't have access to Google as I do. (He was, after all, Newton!) I'd like to conduct a test "for the common people" for the universality of G. The idea is that the centrifugal force of the rotation must always balance out the centripetal force of gravitation.

First we see that in the law relating to gravitational forces we can cancel out the mass of the accelerated body:

$Gm_1 \times a = G \times m_1 \times m_2 / R^2$ meaning where $a = m_2 / R^2$.

This is exactly what Galileo tested at the Leaning Tower of Pisa when he let two bodies of different weights drop. This shortcut gives us the advantage of not having to worry about the mass of the moon and kinds of apples. We can take any apple we please to calibrate the acceleration caused by the earth's gravity. The earth's gravitational acceleration is exactly that speed a body reaches in a unit of time when it is in freefall to the earth's surface. According to Newton's law (I), this acceleration is just that gravitational force that influences the standard mass, which according to the last formula is:

$G \times$ earth's mass/(earth's radius)^2.

With a good stopwatch (I don't know how good Newton's was), after a second's time, every freely falling body reaches a speed of about 10 m/s. This means that the acceleration due to earth's gravity on its surface is simply ten of our units. According to (II) the earth's gravity shrinks by (earth's radius/distance to moon)^2 on the moon, a factor of approximately 10^{-4}. There it has to keep the moon in orbit, but how great is the moon's acceleration when it orbits the earth? To find this out, we need the changes in the moon's velocity per unit of time. In one week, the moon covers a distance almost equal to the distance between the moon and the earth, so that its velocity approximately = (the distance to the moon)/week. Over the course of two weeks it has reversed its velocity while it orbits the earth. This shows its acceleration. To get numbers, I would therefore offer the compromise that the moon's acceleration = distance to the moon/(week)^2. For our test we need to convert these quantities into our units m and s. Today, we know the distance to the moon down to the last centimeter by measuring the time taken by a laser beam reflected by the moon's surface, but we don't need to be that precise. The light takes about a second to get to the moon, meaning that in our units:

distance to the moon = speed of light c times its flight time. c is known to be $10^{8.5}$ m/s, and the second factor is just 1.

One week = 7 days ~ $10^{5.75}$ s. Counting together the acceleration of the moon on its orbit (in m/s^2):

distance to the moon/(week)^2 = $10^{8.5}/(10^{5.75})^2$ = $10^{8.5-11.5}$ = 10^{-3} = 10×10^{-4}.

The last figure is the value 10 of the earth's gravitational force on its surface reduced by the ratio of the distances to the earth's center and to the moon squared $(10^{-2})^2$ = 10^{-4}. According to (II) this is just the force that binds the moon to the earth, and this proves that Newton was right. Of course, I've rounded off some of the figures so that our equation comes out evenly. For a more precise test we'd have first to learn differential calculus, but for the extrapolation heaven-earth, the result is more than adequate.

Newton's vision was a defining point in intellectual history and cries out to be dreamed further. Couldn't we build on the conclusion from the apple to the moon, and extend the argument from the moon to the sun, from the sun farther up to the middle of the Milky Way? It's not meant to be hard. The thought process remains the same, just the powers of 10 expand a bit. Another question is whether we humble creatures should dare to dictate how heavenly bodies are to move over the expanse of the universe. Right now we're exhausted from the calculations with the moon, but in appendix 8 we will dare, and we will succeed! The sun's gravitation is more or less equal to the centrifugal force of the earth's orbit. Similarly, the gravitation of the center of the Milky Way binds the spiral arms so that they can't escape. Actually, in the universe there has to be another dark matter lurking around. We just see how its gravitation moves the other stars; the visible matter is not enough.

We can rest our conclusions on a somewhat more solid basis by observing the situation in a co-rotating frame of reference. For example, in the solar system, the earth would be standing still, because its distance to the sun changes only by a few percentage points over a year. In a rotating system you can feel a centrifugal force, and it has to neutralize the gravitational pull. This force is distance/(orbital period)^2, so that when we check on the compensation of forces, we get the same equivalencies as before:

distance/(orbital period)^2 = $G \times$ sun's mass/(distance)^2.

We'll still want to reformulate these conditions a little so that they become an easily understandable universal statement. When we multiply by (distance)^2 and (orbital period)^2, then the denominators disappear and the equation is:

$G \times$ sun's mass $\times$ (orbital period)^2 = (distance)^3.

Or to break it down for the orbital period:

Orbital period = γ(distance)$^{3/2}$ (III).

Here

$\gamma = (G \times$ sun's mass$)^{-1/2}$

is equal for all planets and (III) is called Kepler's third law. It states that the outer planets are the slower ones, because their speed ~ distance/orbital period ~ (distance)$^{-1/2}$.

> Brain jogging exercise: With his third law, Kepler would have been able to determine the time Sputnik needed to orbit the earth. Try and figure it out yourself.

> Hint: earth's radius = 6,000 km, distance earth-moon = 380,000 km

> Solution: According to Kepler (III) Sputnik's orbital period is:
> orbital period of the moon $\times$ (earth's radius/distance earth-moon)$^{3/2}$.

Now let's cheat a little and say that 6,000 km/380,000 km = 6/384 = 1/64 belongs in the parenthesis above, because that makes it easier to find the roots, ( )$^{1/2}$ = 1/8, and ( )$^{3/2}$ = 1/512, which we can round off to 1/500. If we calculate the moon's orbital period generously at thirty 25-hour days, we get 750 hours, and Sputnik needs 750/500 hours = (1 + ½) hours.

We were generous in our calculations, because Sputnik didn't fly exactly along the earth's surface (it even took a few minutes less). It's just nice to know that we can find our way around in the universe without higher math or computers.

By the way, this time period of a little over an hour has a universal significance. In Kepler's third law, the mass of an attracting body is divided by R^3, so only the density $\rho = MR^{-3}$ occurs. So it predicts that the orbital period of rotations that are almost touching around all bodies with the same density, but with different R, is the same. The equation can be written:

$$G \times (\rho R^3) \times (\text{orbital period})^2 = R^3,$$

meaning that for every R:

$$\text{orbital period} = (G \times \rho)^{-1/2}.$$

The practical effects of the universality of gravity are:

a) In an extremely slow waltz where it takes over an hour to complete one turn, it's not even necessary to hold your partner as long as you're close enough—gravitation will do it for you.

b) At home, the Little Prince by Saint-Exupéry can move a little, but not very fast. Let's say that his little asteroid has the typical diameter of 1 km, making its circumference about 3 km and it's made of the same material as the earth. This means that it would again take about an hour and a half to complete a close orbit. The Little Prince would then only be able to take his evening constitutional at a speed of 2 km/h or else the ground will slip out from under his feet.

Whoever doesn't like the roots from ( ) in (III) and is able just barely to tolerate the complexity of $(100)^{1/2} = 10$ can find a good illustration of (III) in our solar system in the innermost planet Mercury and the outermost, Pluto. Their distances to the sun stand in relation to each other like 1:100 and their orbital periods like 1:1000. (III) requires

> 1000 = orbital period Pluto/orbital period Mercury
>
> = (distance to the sun Pluto/distance to the sun Mercury)$^{3/2}$
>
> = $(100)^{3/2} = 10^3 = 1000,$

and that's exactly what it gets.

Kepler (III) is nicely illustrated in our solar system in its universal form, too. There are the so-called double asteroids, which are two rocks no larger than a kilometer in diameter that almost touch as they circle each other. Their orbital period has in fact been observed to be about one and a half hours. Speaking of Pluto, let's end this section with a last bit of family gossip about the escapades of our youngest sibling. Its big (and somewhat precocious) brothers have lots of funny and frightening things to tell us.

Neptune: Someone should really teach our littlest brother some manners—an orbit that deviates 30 percent from a circle is just not done.

Uranus: All of our orbits are elliptical by a few percentage points. Kepler got himself into a fine mess when he postulated in his first law that all planetary orbits are elliptical.

Neptune: A circle is certainly more perfect than an ellipse. Maybe people got so attached to circular motion because the fixed stars turn in such fine circles in the night sky.

Uranus: People should have been able to figure out from Newton's equations that at least the speeds, if not the locations, of the planets move according to perfect circles. Then they'd have their perfect circles.

Neptune: That's not only true for planets, but for comets as well. But it's only in circular orbits that the velocity circle is centered around zero in velocity space.

Uranus: It would have hurt the earthlings' pride to be on the same level with these cosmic vagabonds. It was good that Kepler kept his mouth shut.

Neptune: Then everyone has kept his mouth shut, because the velocity circle is largely unknown.

Uranus: I read about it in something by some guy called Feynman, but then his heirs got into an argument about the manuscript and the lecture got lost.

Neptune: I saw it in a book about mathematical physics, but it was buried between so much mathematical gobbledygook that it will surely remain our secret.

Uranus: That's just as well, as you can't tell how some fanatics would react.

Neptune: Even if he goes in velocity circles, it bothers me that this little upstart Pluto always is running around between my legs.

Uranus: You mean he cuts across your orbit.

Neptune: Exactly. It's just up to now I was never right there when he did it—sooner or later, though, this reckless behavior will end in a catastrophe.

Uranus: You mean a collision?

Neptune: Then the little shrimp would just completely crash down into me and there would only be eight of us left.

Uranus: But even in a near miss, he'd be so thrown off course that he'd either be catapulted into the universe or fly toward the sun and mess up other brothers.

Neptune: It doesn't bear thinking about, all the havoc the little guy could wreak. You don't have to jump immediately to the worst conclusion, a collision. Even if he were to get too close to the earth, the tidal forces would shake up the continental plates to such an extent that the earth would be deformed and its interior would seep out.

Uranus: That's what happened earlier when some planet gone wild ripped the moon out of the earth. That must have been a terrible catastrophe, and yet it's bound to happen in about a billion years. Objects like Pluto have a period of about 10^3 years, so in 10^9 years, the orbit returns 10^6 times. If the orbit uniformly covers a circle with a radius (distance to Pluto) of 10^9 km, then it comes at least within $10^9/10^6 = 10^3$ km to each point inside. Thus once in a while it would be 300 times closer than the moon.

Neptune: The lunar catastrophe turned out to be a blessing in disguise. The moon stabilized the earth's axis; without the moon, the geographic poles would be wandering around the earth's surface.

Uranus: There must have been lots of wild planets earlier. These rowdies were eliminated slowly over time, and our solar system became an oasis of peace.

Neptune: Finally there's no one getting in someone else's way, but our boat is full now. The way to get here was something like biological evolution, pure Darwin. In our case, though, the most peaceful survived.

Uranus: And here people were thinking that our family was something like a clock that a higher being had set in such a way for us to be able to have a life-giving sister, the earth. We didn't need this omnipotent intervention. Our evolution was completely inevitable.

Neptune: There still could be a plan behind it, though. The time it took from the initial chaos until peace reigned had to be coordinated with biological evolution so that life could originate.

Uranus: I couldn't care less about that. The others should adapt to us, not the other way around.

Neptune: We can't afford to have that attitude if the universe were to collapse again before we have orderly conditions. All of these developments take billions of years. Let's turn back to the present. Pluto's the only one who can disturb things now.

Uranus: Well, there's not going to be a catastrophe any time soon. Although not too long ago this little imp made a fool of all honest astronomers.

Neptune: How's that?

Uranus: You remember back when astronomers couldn't see very well? The only way they saw that we exist was because we disturb the orbits of our brothers a little. A Percival Lowell made some calculations using our orbits and pointed out where another rowdy had to be hiding. That's exactly where the astronomers looked and discovered Pluto.

Neptune: Don't make me laugh. He has just one-thousandth of our mass and is supposed to be able to mess up our orbits?

Uranus: That's what people didn't know at first. It was only over time that it became apparent what a weakling Pluto was, and this triumph of astronomy turned into an embarrassment. They postulated a much too large mass for Pluto, and it was pure coincidence that he was actually found in the foreseen place.

Neptune: It turns out that he's actually something of a freak, almost like a Siamese twin. He's a planet with a moon of almost the same size, and they tightly orbit around each other.

Uranus: There must have been some reason for irregularities in our orbits.

Neptune: Maybe because there are all kinds of waste products floating around beyond Pluto. Just small fries, but already eighty of them have been found.

Uranus: I wouldn't call them brothers, but, after all, even a pebble causes a ripple in the ocean.

P.S. In the meantime, even the astronomers were embarrassed by Pluto and took him off the list of planets.

THE PLANETARY SWING

When taking the mutual attraction of the planets into account, even a Newton is rendered powerless.

Johannes Kepler considered the elliptical orbits of two bodies attracted to each other by gravity to be a reflection of the harmony of the spheres. It just takes one more body, though, and chaos can break out. The motion equations no longer are integrable, and only a few discernible phenomena appear in the sea of the unknown. Simplicity can dominate when there's perfect symmetry. If the three bodies form an equilateral triangle, then there's an answer to the motion equation that says that each body pulls on the common center of gravity of its Kepler ellipse, but in such a rhythm that the triangles always remain similar at different times. The motion remains at the level of the triangle. If the third, lighter body moves perpendicularly to the elliptical level of the other two bodies' orbit, then almost anything can happen. To put it in mathematically precise terms, for every increasing time sequence $t_1 < t_2 < t_3 \ldots$ with a certain minimum distance d between the times, there will be an orbit that crosses the elliptical level at the times $t_1, t_2, t_3. \ldots$

These are artificial constructs and do not reflect typical behavior, which would be that one of the bodies breaks out and moves on. This is possible if, from the very beginning, it flew by the other two at a great distance, so that the contact remains fleeting. Or it could get in their way to such an extent that they fly by one another so closely that one of them gets enough energy that it takes its leave. This can be prevented by saying that everything is actually moving on a round surface, so that if one wants to escape to the east, it will show up again in the west. In this situation, the motion becomes wilder and wilder and heats up. If there are more than three bodies, it is also possible for them to stoke their motion so much in endless space that one of them, after a finite period of time, will reach spatial infinity. When pressed for an explanation as to what it will exactly do when it gets there, the mathematician remains silent and the physicist starts to stammer. After listening to such horror stories, which are not a product of my imagination but illustrate mathematical theorems, we should consider how it's possible that such peace and quiet have settled in our solar system.

To begin with, the forces the planets exert on each other are much weaker than the sun's gravitation. Gravitation is proportional to mass, and therefore that of the sun is thousands of times greater than that of the largest planet. The distances between the planets are also larger than their distance to the sun, so that the force of the sun is typically 10^4 times stronger. This still offers no explanation as to why the solar system has remained stable for billions of years. If we were to stop Mars and Jupiter and if they were alone in the universe, they would crash into each other in a few hundred years. The planetary workings function so perfectly because the disturbing forces are constantly changing direction and become neutralized over time.

If you were only to observe the distance to the sun from an undisturbed elliptical orbit of a planet, let's say the earth, then this is a purely periodic process. Within a half-year the earth gets approximately 1 percent closer to the sun, then turns around and reaches its maximum distance away from the sun after another half-year. Consider what kind of influence Jupiter could have here. It has the greatest effect when it's on the same side of the sun as the earth (conjunction). In this case, Jupiter pulls the earth a bit outside. The earth is faster than Jupiter and can't be caught; it slips over to the other side of the sun. The same thing happens again the next year—it's like when the Grandpa gives his grandchild a push in a swing every once in a while, except it doesn't have the exact same rhythm as a swing; Jupiter isn't standing still. The next push (better: pull) takes place a little later, perhaps in more than a year, and we know how the motion will look from experience at the playground. In the beginning the pushes will build up, but if Grandpa isn't paying attention and misses a beat, there will be a counter-push and the swing will come to rest. Then the whole game begins again, but seen over a long period of time a periodic movement slips in without resulting in a catastrophe. This is indicated in Figure 5.1 for various initial distances.

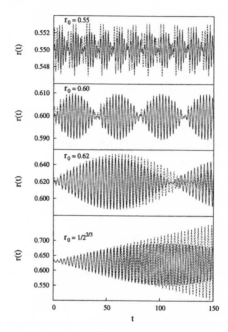

Figure 5.1: Radial deformation of a circular orbit of a light planet with the initial radius r_0 caused by Jupiter. $r(t)$ represents the disturbance in the planet's distance to the sun, with $r(0) = r_0$. In the units used here the distance Jupiter-sun is 1, Earth-sun 0.19, and Mars-sun 0.28. Jupiter's mass is given as 1/999 of the sun's mass. The extended curves are the result of a computer simulation; the dotted line curves are taken from a simple oscillatory model.

A catastrophe would only happen where there is what is called resonance, meaning when one Jupiter year would be exactly two Earth years long, for example. Then, at the next conjunction in two years, Jupiter's force would hit in exactly the same spot. (See the following diagram where S = sun, E = Earth, J= Jupiter.)

Year 0	Year ½	Year 1	Year 1 + ½	Year 2
	J			
S E J	E S	J S E	E S	S E J
			J	

The heavenly clockmaker thought of something for this worst-case scenario as well. First of all, with resonance, the deviation of the earth's orbit from a circular orbit would swing out wider and wider until it loses the rhythm. Even though Jupiter continues to go around in circles, a stronger deformation of the earth's orbit would result in a change in the orbital period. This means other counter-pushes and the old balance returns (Figure 5.1). Resonance (almost) exists in our solar system, not 1:2, but 2:5 between Jupiter and Saturn. In Figure 5.2 we can see the distance between Saturn and the sun over time. Because this is a higher resonance, there is a more complicated back and forth, but no catastrophe. This does not occur in our planetary family because the following rules for peaceful coexistence are upheld:

1. All orbits are practically on one level.
2. The mass of the largest planets is still less than a thousandth of the sun's mass, so that the sun's dominance remains unthreatened.
3. The planets keep their distance from one another and their orbits don't cross (with the exception of Neptune and Pluto).

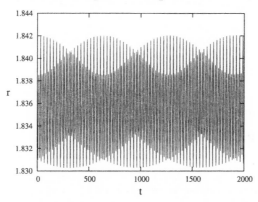

Figure 5.2: Radial deformation of the Saturn's circular orbit with a radius of r_0 $(2/5)^{-2/3} \approx 1.8420$ caused by Jupiter. $r(t)$ represents the disturbance in the planet's distance to the sun, with the distance Jupiter-sun equal to 1. Jupiter's mass is given as $1/999$ of the sun's mass. The curves are the result of a computer simulation.

Things will come to a bad end if one of these rules is broken. In Figure 5.3 we see what would happen if rule 2 were broken and the mass of Jupiter were to be increased from $10^{-3} M_0$ to $10^{-1} M_0$ (M_0 = sun's mass). Then anything from a poor little asteroid to an earth-sized planet would be thrown out of the solar system after almost one hundred Jupiter years. In the same situation, rule 3 is also broken under Jupiter's influence. When planetary orbits cross one another, a catastrophe will occur sooner or later.

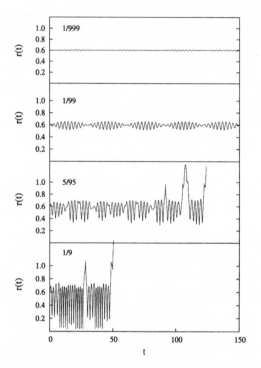

Figure 5.3: Radial deformation of the presumed circular planetary orbit with a radius of r_o= 0.6 caused by Jupiter. $r(t)$ represents the disturbance in the planet's distance to the sun, with the distance Jupiter-sun equal to 1. Jupiter's mass increases from top to bottom, from 1/999 to 1/9 of the sun's mass. The curves are the result of a computer simulation.

This raises the question as to who taught our planets the rules they need for peaceful coexistence? Much can be understood according to Darwinism; the troublemakers were exterminated without exception. There were surely many more planets in the beginning, and there were collisions among them. The asteroids between Mars and Jupiter are fragments of what survived catastrophe. Some people think that our dear moon is also just a fragment like this. It's difficult to say how likely it is for peace to come about after that kind of slaughter. The best way to dispose of troublemakers is to throw them out of the solar system. As they say, out of sight, out of mind. How it's possible to transform planetary junk into galactic junk is the subject of negotiations between a good student, a wise guy, and an eager professor.

The Beginner's Guide to the Galaxy

Two students are talking excitedly after their astronomy class:

Good Student: It sure is scary to think about how, in the beginning, one planet just threw another out of our solar system.

Wise Guy: And to think that they even could—they're just dust specks compared to the sun!

Good Student: They would have to have instincts like the cuckoo, which also throws out other baby birds from the nest.

Wise Guy: They don't have any instincts at all; they can only do what Newton lets them.

Good Student: We should go to Astronomy 101. The professor is supposed to explain how you can leave our solar system.

They cram themselves into the overflowing classroom and cheekily take their seats in the first row.

Professor: Ladies and gentlemen! Our topic today is the survival of our species or even life itself in the distant future. As you know, the sun will expand into a red giant in a few billion years. This will turn our oceans into steam, and we'll have to find a new place to live.

Wise Guy (to Good Student): This guy just wants to pack the auditorium with windbags like himself with his philosophizing. I want something I can get my teeth into.

Good Student (to the Professor): I thought that when the sun uses up all of its nuclear fuel it would simply cool down.

Professor: This is where the sophisticated effect of negative specific heat comes in. According to it, energy loss/warming/contraction all belong together and correspondingly energy gain/cooling/expansion. When the hydrogen fuel is used up, the sun's core contracts; it becomes hotter and releases energy. The outer layers absorb this energy and cool down and expand because of negative specific heat.

Good Student: So it does cool off.

Professor: Yes, but also gets so big that it reaches the earth's orbit, and it's simply too much to have a thousand degrees outside your door. This is a 101 course, and I'm not assuming that you're familiar with dynamic system theories, computer algorithms, and other advanced knowledge. You've just got to know the following planetary rules of the road:

For a satellite with the standard mass:
1. Kinetic energy = $E_{kin} = v^2/2$, where v is the velocity
2. Gravitational energy = $E_{grav} = -GM_\odot/R$, where $M_\odot$ is the sun's mass, and R the distance to the sun. We are only taking its effect into account.
3. For a circular orbit around the sun $E_{kin} = -1/2E_{grav}$ (virial theorem).
4. An orbit exits our solar system exactly when the total energy is positive: $E = E_{kin} + E_{grav} > 0$.

To further elaborate the following points:

re 1. The kinetic energy sharply rises with increasing velocity v.

re 2. The gravitational energy, on the other hand, becomes more negative the shorter the distance to the sun becomes.

re 3. The negative gravitational energy for circular orbits is halfway balanced out:

$$E = 1/2E_{grav} = -E_{kin}.$$

re 4. Regardless of how else the orbits are, when $E > 0$, it will ultimately find its way out of the solar system. If $E < 0$, it will always be held captive.

Good Student: I've heard that before, but it's been so long ago that I don't remember where or when.

Professor: To be able to exit our solar system starting from the earth, the kinetic energy must be increased so that $E > 0$. To this end, let's give the satellite a velocity of Δ. From the universe's point of view, this is in addition to the earth's velocity v_E around the sun (about 30 kms). When they have the same direction, then in keeping with (1), the satellite's kinetic energy will be $(v_E + \Delta)^2/2$, while its gravitational energy will remain the same. If NASA could reach $\Delta = v_E$ and double v_E, then we'd already make it out:

$$E = (v_E + \Delta)^2/2 - GM_\odot/R_E = 4v^2_E/2 - GM_\odot/R_E = v^2_E > 0,$$

because (III) demands that $v^2_E = GM_\odot/R_E$. We can find out the speed at which we race into space by considering the following:

far away from the sun the energy $v^2/2 - M_\odot G/R$ becomes almost $v^2/2$, as soon as R is sufficiently large on the journey out. The total energy remains v^2_E, however; therefore the velocity v_∞ far away must be:

$$v_\infty = v_E\sqrt{2} = 1.4v_E.$$

Wise Guy: But teacher . . .

Good Student (nudging Wise Guy): Professor!

Wise Guy: Professor, I would like to do it the other way around and have $\Delta = -v_E$.

Professor: Are you out of your mind?

Wise Guy: What you just said about negative specific heat sounded so crazy that it seems to me the best thing would be to put on the brakes in order to accelerate.

Professor: But then the satellite would be standing still in the universe, and the sun's attraction would make the satellite crash into it.

Wise Guy: If I were on the satellite, I'd stop it right when I'd be at the point where I'm falling toward the sun. I'll just miss Mercury, and by orbiting Mercury it will reverse my velocity and get me out of there.

Professor: And what could you possibly win by this devil-may-care slight of hand?

Wise Guy: Quite a lot in fact! I'm thinking it will be like this: Mercury seems to reflect by close circumvention. After reflecting on the moving

Mercury, I would have twice its velocity and therefore the same situation you had in your demonstration (Figure 5.4). The only difference is that I have, far out in the universe, 1.4 times the velocity of Mercury, and that's much faster than the velocity of the earth. It even gets better because then there's still the velocity of the fall to the sun.

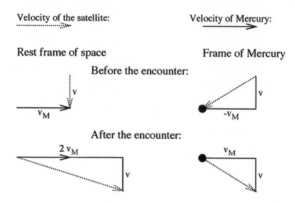

Figure 5.4: Encounter between a satellite and Mercury as seen from the system in which the universe is at rest (universe system) and the system in which Mercury is at rest (Mercury's system).

I'd like to present this scenario as an exercise to do as homework. This results in the velocity v_∞ after leaving the solar system being (see appendix 9)

$$v_\infty = 5v_E$$

Good Student: That's great. Your initial investment on velocity was Δ = 30 km/s, the same amount of energy as in the professor's rote method. But you generated three times the velocity for v_∞. Maybe you'll even be able to get a job at NASA when you're saving Uncle Sam so much money anyway. They probably know that already; I've even heard on TV that other planets can give you impetus.

On the way home:

Good Student: Still, galactic travel isn't my cup of tea. Even if we generate v_∞ = 300 km/s, 10^{-3} the velocity of light, the nearest useful star would still be ten light years away and the flight would take about 10^4 years. I think the whole thing would get pretty old in ten thousand years.

Wise Guy: Perhaps this is not insurmountable. Suppose for comfort's sake, you keep accelerating in the same direction with the earth's acceleration. Then, in about a year, you'd come close to the velocity of light, and gain a time dilation factor. This you can use so that you can reach any point in our galaxy in another year of proper time.

Good Student: Two years isn't so bad. After all, people have gotten used to such travel times in planetary travel.

Wise Guy: The serious problem will be to find a sponsor for the gigantic energy bill. Even without futurology, we could see how easy it is for two planets to kick each other out of the solar system. The fact that Mercury has a much smaller mass than the sun didn't even play a role in my theory. Mercury's mass needs to be large enough so that its force overpowers the force of the sun when it's closely orbited. Mercury also must be heavier than the satellite.

Good Student: Sure, because if both had the same weight, then one couldn't dictate the resulting flight direction, and in the end both might crash into the sun. People were right for thinking that someone has to watch out in planetary traffic.

This conversation shows how events in Newtonian mechanics become many-faceted as soon as more than one player is involved. There are even circumstances that the Wise Guy overlooked. If you want the maximum kick from Mercury, you have to get so close that you will touch its surface and be gone. It's often said that these problems have already been solved in principle. This is true, but what we're concerned with is located in the inexhaustible number of isolated cases.

6

Why Does Life Exist?

WHAT IS LIFE?

*Chemical forces are able to create the most astounding things
out of atoms as if by magic. Life continues this process
and takes it to the extreme.*

A VISITOR from another star would have to be impressed by the sheer
variety of surfaces of our inner planets—Mercury: blackened lava;
Venus: white clouds; Mars: red deserts; Earth: blue seas. On closer inspec-
tion the green shimmer of our continents would start to be apparent, and
our visitor's astonishment would only grow with each step closer. It's not
only chemistry that's been at work. A much more extensive principle created an apparently inexhaustible biological variety of species. Just the number of molecules that join together from the four elements H, C, N, and
O is enormous, but, from the cosmic point of view, an exception. Most
visible matter is found in the stars, and there is only one material to be
found inside, so-called jellium, which is a soup of pretty much evenly distributed electrons in which the atomic nuclei are swimming around—
mainly hydrogen nuclei or protons, along with 25 percent helium nuclei
(α-particles) more or less seasoned with some heavier atomic nuclei
depending on the star's age. The characteristics of this standardized soup
are dependent on its temperature and density; it has no other structures
or specifics. Only after the pressure of gravity has been reduced does chemistry come into play and the variety of molecules begin to flourish. Why
do these come together to create the enormous number of living creatures

that overrun our earth? Our extraterrestrial visitor would be at pains to guess why. The traces of our origins have been scattered like those of what happened within the first three minutes after the big bang. It wouldn't be possible for us to develop today; our atmosphere now contains too much oxygen. Originally a byproduct of biological evolution, it became poison, an addictive poison. Although it is lethal for many bacteria, we can't survive without oxygen for more than six minutes.

Even though it hasn't yet been possible to re-create in a retort how life began, the number of human attempts to explain the origins of life is almost as vast as that of biological species. Here, I can naturally only offer a few samples and touch on some topics dealt with in two influential books of the twentieth century, *What Is Life?* by Erwin Schrödinger and *Chance and Necessity* by Jacques Monod.

One of the few responsibilities Schrödinger had in Dublin was to give a few open lectures for the general public each year. He usually then developed his notes into short books published by Cambridge University Press. They are all excellent examples of his brilliant style, but none had as much of an impact as *What Is Life?* In it he deals with two central questions.

1. Life is doubtless a kind of order which stands out from thermal uniform distribution. But doesn't this contradict the second law of thermodynamics, according to which entropy, and therefore disorder, is constantly increasing?

2. What has given the genetic makeup such stability that it can be handed down through generations and generations in its unadulterated form?

From what we know today, he already was able to recognize the right answers.

re 1: The second law only demands that total entropy increase and that it's only possible to view the living system in its interaction with its environment. When the entropy sharply increases in the environment, then the entropy of the living system can decrease to a certain extent, although the total entropy, the sum of both, increases.

re 2: According to quantum theory, there are, in contrast to classic theory, structures that are fixed and do not constantly go through small changes over time. Schrödinger thought that genetic material could be similar to an aperiodic crystal: aperiodic because the genetic information has to be transmitted by something like Morse code, and crystal so that it doesn't quickly become deformed.

Schrödinger's book got a whole young generation of physicists interested in these questions and caused a developmental leap in molecular biology. This didn't prevent open criticism of his ideas, however. On the one hand he was accused of having nothing new to say. This was just empty criticism because the point of a popular lecture is to make things clear and accessible to the general public, not to announce something new. More serious was the accusation that his priorities were wrong; the increase in entropy wasn't such a central problem. In an open system like one for a living creature in its environment, entropy doesn't have to increase, but rather free energy (= energy – temperature × entropy) should decrease. The first energy part is more important; after all, we eat energy-rich sausages and not entropy-poor diamonds. In reality, we earthlings get a lot of entropy-poor energy through sunshine. We get photons of a kind of energy from the sun, which corresponds to the 6,000° K of the sun's surface, and each photon is split into 20 photons with the earth's temperature of 300° K. This is demanded by the conservation of energy, which is proportional to temperature, and 20 × 300 = 6,000. The entropy of a photon gas is basically the number of photons, which by this split is increased by a factor of 20 on earth. At their expense, we're able to afford quite a bit of biological order without reducing the total entropy. How much is dependent on some details. I once estimated that in one summer it would be possible to afford a 10-m-high forest from the sunshine without crossing the principle of increase in entropy (the second law of thermodynamics). A colleague in California told me that he repeated these calculations and came up with a height of 20 m. I could only answer that they probably had better summers in California. In any case, the question of increasing entropy is a psychological barrier for physicists, and Schrödinger was right on track to have discussed it.

Regarding the second point, it was said that an aperiodic crystal was a contradiction in terms, as per definition a crystal is something periodic. When the molecular reproduction mechanism, DNA, was decoded a decade later, it was clear that Schrödinger's intuition wasn't far off the mark. DNA is a very long molecular string on which four pairs of the basic construction blocks—the four bases adenine, guanine, cytosine, and thymine—are strung together. A pairs up with T and C with G. When the pairs are written vertically and the string is lying horizontally it can appear like:

```
A   T   G   C...
|   |   |   |
T   A   C   G...
```

All of the information is carried in the order of appearance. The string is twisted into a spiral with the appealing name "double helix," but this is irrelevant for the reproduction mechanism. This functions quite simply—you pull it apart, like unzipping a zipper:

```
A   T   G   C...
|   |   |   |
T   A   C   G...
```

and because everything's swimming around in a soup of the four bases, every base is looking for its appropriate partner:

```
A   T   G   C...
|   |   |   |
T   A   C   G...
```

```
A   T   G   C...
|   |   |   |
T   A   C   G...
```

Then the strings split up; every string goes its own way, and we've already been cloned:

```
A   T   G   C...
|   |   |   |
T   A   C   G...
```

```
        A   T   G   C...
        |   |   |   |
        T   A   C   G...
```

I still consider this process to be a miracle, as I've had plenty of experience trying to repair chain-link fences. To do so, two wire spirals have to mesh together perfectly. They never did by themselves; rather there was always a hopeless tangle of wire. Of course, the biochemists say that quantum mechanical binding forces are at work here, but the fact that they are able to create more order never ceases to amaze me. The following thought can point us in the right direction: the secrets of life are hidden behind quantum phenomena, which are not directly accessible to human reasoning trained to deal with ordinary day-to-day events. The molecules must press up against one another tightly so that the forces affecting short distances start to play a role. These forces are a wonderful compromise. They must have a shorter range than the electrostatic forces, but a longer range than the main molecular binding forces. They must be strong enough to keep the molecules together, but not too strong so that it will still be possible to unzip the zipper. They are called van der Waals forces and have been known about for a long time, but they have their source in the correlations between the molecules that only quantum mechanics provide. Today we say that the molecules must be entangled. Schrödinger originally used the rather awkward German word *verschränkt* (crossed or twisted) to describe this. It's no accident that Schrödinger, a genius at languages, couldn't find a better word for it. This feature can't be expressed in everyday language because everyday objects aren't "entangled" in any obvious way.

Monod's book appeared in 1970, five years after he won the Nobel Prize in medicine. It proves at first glance that the author is actually a chemist. The book describes the reproduction mechanism in great detail. At the outset I called this book dogmatic. It places the dogma of objectivity over everything. The question is what exactly does this mean. It is subject to some dispute in quantum mechanics in its naïve form, "Every statement must be independent from its observer." The principle of objectivity must mean something, but until it's worked out exactly, it appears to be arbitrary when it can be used to extinguish possible new approaches. Irrespective of all that, the book offers an overwhelming panorama of biological evolution. As the title says, it investigates what follows as a matter of course from the first principles and what has a random source. Monod reaches a pessimistic conclusion: "Human beings are products of chance, infinitely alone in the vastness of the universe, perhaps only a tiny spark in cosmic evolution." How much more optimistic the vision of the proponent of the eschatological anthropic principle discussed in the final chapter: "The universe was only created to one day be at the service of human beings." To reach this state of affairs , unbelievable amounts of blood, sweat, and tears will have to be shed, and in the meantime we will have to be satisfied with some shorter-term goals.

First, we should look at necessity, strictly deterministic laws. We will see how increasing entropy or survival of the fittest are derived not only verbally, but also mathematically-formally. Don't worry: we will limit our studies to embryonic forms that don't require more than basic math. I hope to couch it in easy-to-understand terms.

In the last section we will take a closer look at one of the many random factors in biological evolution—namely, water. It not only gives our planet its blue majesty; it is a miracle from whichever way you look at it. Its melting point, boiling point, density, and dielectrical constant, among many other features, are exceptional compared to other substances containing hydrogen H_2X. Why? There's one small difference: in other molecules with the form H_2X, the hydrogen branches form a more or less right angle; with H_2O it's wider, 104°. This nondescript detail has serious consequences that are crucial to the development of life.

CREATION OF ORDER

Why is disorder always emerging even when the development
can take place in the reverse time order and could create order?

The increase in disorder over time seems to be a general principle. In physics, it appears as increasing entropy and can be seen in almost every branch of science. Biology seems to be the exception, as life forms are highly ordered systems. It's first of all necessary to explain why life forms defy the second law. We've already seen in chapter 4, pages 75–86, that the general striving toward even distribution can allow a partial system to become more orderly. With the movement of mutually attracting bodies, they build spatial groupings. Of course, this means more order as a completely homogenous distribution. The increase in this order was more than balanced out by a simultaneous increase in disorder of their velocities caused by heat. When the distribution of points was more concentrated, the distribution of velocities was so blown apart that the total entropy ultimately increased.

Naturally, it's hard to keep track of the motion of so many particles, which is why we will study the changes in entropy using a simple example of just three parts. The parts (abstract *a, b, c*) are personified by three players, commonly referred to as Alice, Bob, and Charles. The game itself is not particularly entertaining, but it shows how order comes about and that a decrease in entropy, even in this embryonic form, can be observed. It will also serve to illustrate the mathematical paradigm upon which every deterministic description is based. We'll first get to the heart of the matter and then let the details form a little play.

The system's state will not be described in intangible terms like probabilities, but rather through something tangible like money. Alice, Bob, and Charles are each in possession of some assets, which will be represented by V_a, V_b, and V_c. The temporal development of the system tells us how the amounts of money (in the form of bank accounts) change over the course of a week. The dynamic law is simply a rule of the game and tells us what has happened to V_a, V_b, and V_c in a week. Let's play around a little with the following primitive rule: Alice gives half of her money to Bob, who gives half of his to Charles, who gives half of his to Alice. The total amount of money stays the same. We can use a currency where this value is equal to one. This kind of rule is a special instance of the mathematical notion of a map: it assigns every element of a particular quantity exactly one element of a possibly different quantity. The two quantities in our case are the account balances on a Monday and the following Monday. The first questions are if the initial differences among Alice's, Bob's, and Charles' accounts will be wiped out over the course of many weeks and if the original distribution moves toward the even distribution of 1/3, 1/3, 1/3 over time. Even distribution is a fixed point in the temporal development, meaning that it won't change because of our rule of capital transfer. It is even a so-called attractor, which means that all distributions approach it. This is best illustrated by representing the account balances by a point in a triangle: because the total sum of the accounts is 1, it's enough just to show two accounts. V_b is drawn to the right and V_a upwards. Figure 6.1 shows how the game progressed for different initial incomes. It shows how all of the accounts approach even distribution in a spiral, but not monotonically.

The next problem is related; will the entropy increase in each step due to this temporal development? In our game, we can use the following simple expression for entropy: total assets squared minus the sum of the squares of the individual assets. Using the rule we've made for temporal development, we will actually see an increase in entropy and an attempt toward even distribution. This is shown in Figure 6.2, where the temporal development is shown going to the right and the entropy for several values of the initial accounts is shown going up.

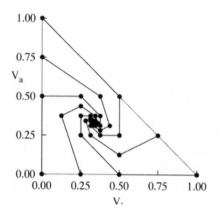

Figure 6.1: Development of Alice's V_a and Bob's V_b bank account balances over time. Charles' account balance is the result of $V_c = 1 - V_a - V_b$. They are all aiming for the attractor.

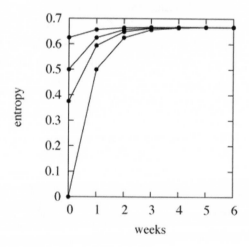

Figure 6.2: Entropy growth over time for different initial distributions in Alice's, Bob's and Charles' account balances. This occurs without variation.

Our game illustrates how entropy increases even though the rule of the game is reversible. One argument against increasing entropy was the so-called time reversal objection (*umkehreinwand*), that temporal development can be reversed. In fact, there is an inverse map for the temporal development that re-creates the initial state when it follows the original map. In our case the inverse is the following simple rule: when the players are sitting in a circle, then each player receives the fortune of the player on one side and must hand it over to player on the other side. This inverse is a different kind of map. It means that the players go deeper and deeper into debt, which leads to instability. It does succeed in decreasing the entropy, even making it negative over time. So it isn't clear if this evolution should be called reversible or not.

Our first rule can be seen as a model for a microscopic law that, at this level, doesn't lead to order, but to even distribution. Nonetheless, from a global viewpoint, this can appear differently. The same game rule leads to a decrease in another, more global entropy. If Alice and Bob combine their lot by getting married, and if entropy is only measured by family income, then this entropy S_f can also decrease according to our rule. This can be seen in Figure 6.3, in which the development of S_f is shown first going up, but then S_f for some initial accounts turns down again.

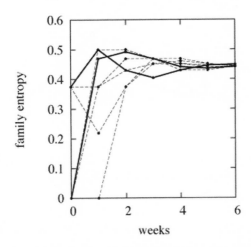

Figure 6.3: Family entropy S_f growth over time for different initial distributions in Alice's + Bob's and Charles' account balances. This development is varied.

This demonstrates that, without contradicting the microscopic laws, macroscopic rules can be formed that exhibit a tendency to create order.

In biological evolution, the fittest survive—that is, the opposite of equal-
ization—and we will also show this situation with our players as a mathematical model. The game rule mentioned up to now is linear (through a
so-called double stochastic matrix), but to illustrate biological systems
we will also use a nonlinear relation. This law says that every player is in
for a certain amount, which speculation then will double. Because the
value-added tax is proportional to the amount involved, the total amount
of money involved in the game retains the same balance. This rule won't
lead to instability, but rather three fixed points—namely, when one of the
three players has all of the money. Only one of these points is an attractor, specifically the one in which the player with the greatest bid gets everything. When we put the account balances in a triangle again, the game
offers the following picture (Figure 6.4).

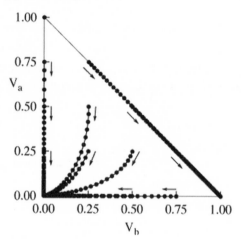

Figure 6.4 Development of the assets V_a, V_b and $V_c = 1 - V_a - V_b$ towards
a fixed point where one player, in this case Charles, will gain them all.

Now everything is circling toward the point where Alice and Bob have
nothing and Charles everything. To carry this analogy over to evolution,
this means that the best one wins. All three points have zero entropy; therefore the entropy decreases while approaching the attractor, but increases
when going away from the other points. This behavior is illustrated in Figure 6.5, where we display the fitness levels. This shows a steady, constant
increase, while entropy gets smaller and smaller in the end.

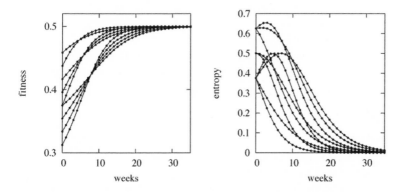

Figure 6.5: The diagram on the left shows the growth in fitness over time for different initial distributions in Alice's, Bob's and Charles' account balances. The diagram on the right depicts how entropy behaves in these conditions.

Of course, our game rules do not presume to ever actually take place in nature. They are just meant to show a scenario of how evolution could occur. The playing of our game requires not only orders of magnitude, but also exact numbers; that's the reason for the fractions. I don't want to force these on anyone, and whoever thinks this will be too much can skip over to the next chapter without missing out on anything crucial. On the other hand, if I've stimulated your curiosity, you will probably find the following entertaining.

Act I

Alice: Nothing new will be created in our distribution game. The total amount $V = V_a + V_b + V_c$ can't ever change in a turn.

Charles: Then I suggest that we create a currency in which our total combined assets = 1:

$$V_a + V_b + V_c = 1.$$

The game will also be easier to play when we just use numbers for V_a, V_b, and V_c to show our account balances.

Bob: Because we're all friends, I'd like to suggest that we start with a game where eventually all differences in income are balanced out. But what's the most just way to do that?

Alice: That's easy: with a communist decree, written as $Co \rightarrow$. According to the decree, Bob gets half of my money, Charles half of Bob's, and I get half of Charles'. In a week the account balances will have gone like this:

Co-decree:
$$(V_a, V_b, V_c) \rightarrow \left(1/2\,(V_a + V_c),\ 1/2\,(V_b + V_a),\ 1/2(V_c + V_b)\right).$$

This should be the rule of our game, our dynamic law.

Director's interruption: What you call decree is actually a central notion in modern mathematics where it is called a map. In its abstract beauty, it just means that you associate an element of a second set to each element of a set.

Bob: Let's see how this will work. How would it be possible to wipe you out financially if you were to have all our money at the beginning?

Alice: According to the decree *Co* the first step would be:

$(1, 0, 0) \rightarrow 1/2, 1/2, 0.$

Charles: I still didn't get anything, let's keep on playing:

$1/2, 1/2, 0 \rightarrow 1/4, 1/2, 1/4.$

Alice: Let's go on to the next round:

$1/4, 1/2, 1/4 \rightarrow 1/4, 3/8, 3/8.$

Now you're richer than I am.

Bob: If we play another round, our accounts will be the same; only Charles will remain a Croesus:

$(1/4, 3/8, 3/8) \rightarrow (5/16, 5/16, 6/16).$

Alice: At some point there should be complete equality (1/3, 1/3, 1/3).

Charles: That'll never happen because the decree says that:

$$V_a - V_b \; \rightarrow \; 1/2(V_c - V_a),$$

$$V_b - V_c \; \rightarrow \; 1/2(V_a - V_b),$$

$$V_c - V_a \; \rightarrow \; 1/2(V_b - V_c),$$

which means that the differences between us will grow smaller, but never simultaneously disappear.

Bob: Unless we had complete communism from the beginning, because:

$$1/3, 1/3, 1/3 \quad \rightarrow \quad 1/3, 1/3, 1/3.$$

Alice: But full equality $(1/3, 1/3, 1/3)$ should happen sometime anyway. If the distribution can't be entirely equal, then we should at least have some indicator to show that it's becoming increasingly fair.

Charles: There's a quantity like that in physics called entropy, and it's written S. In our case this would be:

$$S = V_a \ln(1/V_a) + V_b \ln(1/V_b) + V_c \ln(1/V_c)$$

Entropy is zero, when one person has all the money, and $\ln 3$, when it's evenly distributed. S therefore increases with a fairer distribution.

Bob: What in the world is $\ln$?

Charles: That's the logarithm. You probably learned it in junior high.

Bob: Please don't talk to me about logarithms. I always fell asleep in math, and when it was over, it was like waking up after a nightmare.

Charles: That's okay. Even in math there are different definitions of entropy. How about

$$S = 1 - V_a{}^2 - V_b{}^2 - V_c{}^2.$$

Then it would be zero for $(1, 0, 0)$ and have a maximum value of 2/3 for $(1/3, 1/3, 1/3)$.

Alice: Someone told me that that's called "Tsallis entropy." Let's see if this S increases with our Co. Let's start with $(1, 0, 0)$, then $S = 0$. Using

Co this will become $(1/2, 1/2, 0)$. This means that the entropy has already gotten larger $S = 1 - 1/4 - 1/4 = 1/2$. Using *Co* again, we'll get $(1/4, 1/2, 1/4)$ and that will give us

$$S = 1 - 1/16 (1 + 4 + 1) = 5/8,$$

meaning that S has increased by another 1/8. It works.

Charles: You don't have to grab all the money for yourself right at the beginning because the increase in S also works by any amount (V_a, V_b, V_c).

Alice: How can you tell?

Charles: I'm better at math than you two and can see that after *Co* has worked S becomes:

$$S = 1 - 1/4 \left[(V_a + V_c)^2 + (V_b + V_a)^2 + (V_c + V_b)^2 \right].$$

I can write this so that we can see the increase in entropy directly:

$$S = 1 - V_a^2 - V_b^2 - V_c^2 + 1/4 \left[(V_a - V_c)^2 + (V_b - V_a)^2 + (V_c - V_b)^2 \right],$$

meaning exactly the same S as before plus some positive value.

Alice: Except for $(1/3, 1/3, 1/3)$, then the extra value is zero.

Charles: That's true. With even distribution, S has already reached its maximum value and can't increase any more.

Bob: As far as I can tell, the game is now starting to get boring—even get a little frustrating. We're getting closer and closer to the communist goal $(1/3, 1/3, 1/3)$ without ever getting there.

Alice: If we really want to get there, we could have used the law:

Sco: (V_1, V_2, V_3) will always result in $\left(1/3, 1/3, 1/3\right)$.

This supra-communist game follows all our other rules, too.

Bob: But it would be even more boring because nothing would ever change again after the first turn.

Charles: It would also be another kind of map, because it wouldn't allow a direct reversal. If a new government came to power and wanted to carry out broad restitution, it wouldn't know how, because all initial distributions lead to (1/3, 1/3, 1/3). Let's forget about this brainwave.

Bob: Maybe we should show what we've found out so far in an illustration. We'll draw how the three amounts are becoming more evenly distributed, and underneath, how entropy is increasing at the same time.

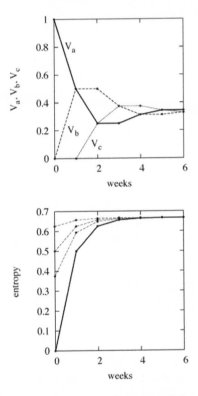

Figure 6.6: The upper diagram depicts the temporal development of Alice's, Bob's and Charles' assets. The lower diagram shows the entropy S growth over time for different initial distributions in (V_a, V_b, V_c). The elongated curve represents the situation discussed by Alice, Bob and Charles (1,0,0).

Charles: I have a better idea. We just need to show two amounts; the third just makes up the difference to 1. If I draw the development of V_a going up and V_b to the right, then every amount will be a point in a triangle. In the course of our game these points wander about (Figure 6.7). We can imagine entropy as being a mountain over the triangle and the movements are always climbing it. This shows that entropy always increases in any turn.

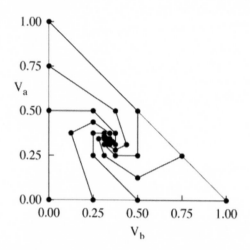

Figure 6.7: Development of Alice's V_a and Bob's V_b bank account balances over time. Charles' account balance is the result of $V_c = 1 - V_a - V_b$. They are all aiming for the attractor.

Alice: All this standardization is getting on my nerves. Let's see if we can find some rules that will at least reverse the effects of *Co*.

Bob: A banker once told me, to make capital, you first have to borrow it from the bank and then cleverly invest it.

Alice: You mean you want to go into debt?

Charles: Going into debt means that V becomes negative, and when we allow that, I can easily make a capitalist rule *Ca* that will reverse the communism:

$$Ca : (V_a, V_b, V_c) \rightarrow (V_a + V_b - V_c, -V_a + V_b + V_c, V_a - V_b + V_c).$$

Alice: Let's see if that works. *Co* made $(1, 0, 0) \rightarrow (1/2, 1/2, 0)$, and *Ca* really does make $(1/2, 1/2, 0) \rightarrow (1, 0, 0)$.

Charles: I see that you've got to have all the money again, but I can show the same thing in more general terms:

$$Co: (V_a, V_b, V_c) \rightarrow 1/2(V_a + V_c), 1/2(V_b + V_a), 1/2(V_c + V_b),$$

which Ca turns into:

$$\big(1/2(V_a + V_c + V_b + V_a - V_c - V_b), 1/2(-V_a - V_c + V_b + V_a + V_c + V_b), 1/2(V_a + V_c - V_b - V_a + V_c + V_b)\big) = (V_a, V_b, V_c).$$

Ca therefore reverses Co at any initial value.

Bob: It's only with perfect communism (1/3, 1/3, 1/3) that capitalism is powerless; even Co can't change anything.

Director's interruption: Bravo! As simple as your map may be, it illustrates a subtle mathematical point. Co and Ca with all iterations exist and form a group of maps. However, since Ca is not positive, we only have a semigroup of positive maps.

Alice: Actually I see Ca as saying $(1/3, 1/3, 1/3) \rightarrow (1/3, 1/3, 1/3)$. Otherwise Ca would have to reduce entropy S.

Charles: In fact, using Ca we have

$$S = 1 - (V_a + V_b - V_c)^2 - (-V_a + V_b + V_c)^2 - (V_a - V_b + V_c)^2$$

$$= 1 - V_a^2 - V_b^2 - V_c^2 - [(V_a - V_b)^2 + (V_b - V_c)^2 + (V_c - V_a)^2].$$

Ca means that the entropy value S will be reduced by the amount in the square brackets!

Bob: This is a fun game. Let's play some more. What happens after Alice has already had all the money once?

Alice: That's simple; $Ca : (1, 0, 0) \rightarrow (1, -1, 1)$, which means that you've got to go into debt and give the money to Charles and I keep my money.

Bob: Hmmm, is capitalism really such a fair system?

Charles: Apropos of being fair, what happens with S—it was already zero at $(1, 0, 0)$?

Alice: *S* becomes 1 - 1 - 1 - 1 = -2. Positivity is ensured for *S* only as long as no one's in debt.

Bob: But where does *Ca* take us afterwards? It seems to me that *Ca*: (1, -1, 1) → (-1, -1, 3), and using *Ca* one more time means (-5, 3, 3). Then I'm suddenly filthy rich and Alice has a mountain of debt.

Charles: In the next round you get even richer at our expense:
    Ca : (-5, 3, 3) → (-5, 11,-5).

and it'll get even worse for me in the next round:
    Ca : (-5, 11,-5) → (11, 11,-21).

Alice: I think we should stop now. We're getting into turbo capitalism, which creates huge inequalities in the distribution of wealth.

Bob: *Ca* just has to. It's supposed to do the opposite of *Co*, after all, and *Co* evens out the distribution of wealth. In our illustration, the transition to *Ca* just means that we're now going from right to left. We could go back over the initial state if we expand the scale of amounts downward. This is suggested in Figure 6.8, which extends the right-hand picture in Figure 6.5 to negative times.

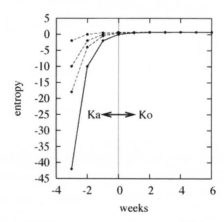

Figure 6.8: Development of entropy using *Co* in a positive temporal direction and using *Ca* in a negative temporal direction. The dotted curves correspond to different account balances of Alice, Bob and Charles at time 0. The elongated curve is the situation discussed in the text (1,0,0).

Act II

In the meantime, Alice and Bob have gotten married, and therefore they view their financial situation from a different perspective. They have combined their wealth into a family income $V_f = V_a + V_b$. Now it's no longer about three people, but rather two families: Alice + Bob and Charles.

Alice: I'd like to see how the communist evolution Co and the capitalist Ca affect family policy.

Charles: What do you think? Which is fairer: when each family gets the same amount of money, or when each family gets an amount proportional to the number of its members?

Bob: I'd like the proportional version, particularly when we have kids someday.

Charles: Why did you get married in the first place then? Well, I don't want that and suggest that we judge the merit of a distribution with the family analogon to the previous entropy.

$$S_f = 1 - V_f^2 - V_c^2$$

Physicists love to call this "coarse-grained entropy" or "entropy reduced to a subsystem." It returns to 0 when one family has got everything and reaches its maximum value of 1/2 with the distribution:

$$(V_f, V_c) = (1/2, 1/2).$$

Bob: I think you're trying to pull a fast one on us. I've read somewhere that coarse-grained entropy always increases so that everything's trying to reach the distribution $(1/2, 1/2)$, which I don't want.

Alice: You shouldn't always believe everything you read. The proof that coarse-grained entropy always increases can't be airtight. Even normal entropy decreases with Ca.

Bob: That's true. When you start with just 1/4 of V_f, and I have the other 3/4, meaning that together we have everything, $V_f = 1$, then Charles has got to go into debt to pay me according to $Ca : (1/4, 3/4, 0) \rightarrow (1, 1/2, -1/2)$.

The family accounts look like $Ca : (1, 0) \rightarrow (3/2, -1/2)$.

This means that the family entropy disappears at the beginning, $S_f = 1 - 1 = 0$. We've got it all, and Ca turns this into:

$$1 - (3/2)^2 - (-1/2)^2 = -3/2$$

meaning that the entropy really is decreasing.

Charles: That even works with Co, without my having to go into debt. Let's start by my having half, Alice the other, and you have nothing. According to Co then

$$Co: \left(1/2, 0, 1/2\right) \rightarrow \left(1/2, 1/4, 1/4\right).$$

The family accounts develop as follows

$$Co: \left(1/2, 1/2\right) \rightarrow \left(3/4, 1/4\right),$$

which means that the family entropy has decreased from

$$1 - \left(1/2\right)^2 - \left(1/2\right)^2 = 1/2 \text{ to } 1 - \left(3/4\right)^2 - \left(1/4\right)^2 = 3/8 < 1/2.$$

Bob: So communism is better for large families after all.

Alice: You don't always have to turn everything into politics. You could also interpret it as saying that, even by observing just a partial aspect, order can be created. That's easy to see in the way you've drawn it.

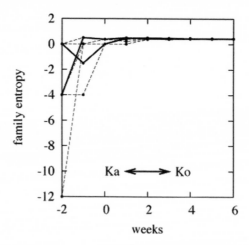

Figure 6.9: Family entropy S_f growth over time for different initial distributions in Alice + Bob's and Charles' account balances. The history of the balances from fig. 6.3 (negative time) is included here.

Charles: But order doesn't have to come about. If Bob, for example, has everything at the beginning, then Co says:

$Co : (0, 1, 0) \rightarrow (0, 1/2, 1/2)$, or by families Co: $(1, 0)$ $(1/2, 1/2)$, or $S_f = 0$ becomes $S_f = 1/2$ after Co.

Family entropy can also increase using Co.

Bob: But that depends on how Alice and I divide V_f. If she got everything at the beginning, then Co would make it Co: $(1, 0, 0) \rightarrow (1/2, 1/2, 0)$; it wouldn't cause V_f to change and S_f would stay 0.

Charles: That shows that Co can't be reduced to a game that transforms one family-style distribution (V_f, V_c) into another. After all, it remains the same in both cases, but Co makes them into something different based on how we've provided the V_f distribution.

Alice: It seems to me that even Ca, which always reduced S, is able to increase S_f. Let's say we have the distribution at the beginning $(3/4, 1/4, 0)$. Under Ca that turns into $(1, -1/2, 1/2)$, so for the family version this would be $(1, 0) \rightarrow (1/2, 1/2)$. This means that the minimum $S_f = 0$ is transformed into the maximum $S_f = 1/2$ using Ca.

Bob: This is getting way over my head, I don't understand what's going on anymore.

Alice: Come on; it's not hard! If Co can be used in both ways, to increase and decrease S_f, then the opposite transformation Ca can do it, too.

Charles: The only lesson to be learned here is that order can also come about when we observe subsystems; this does not contradict the increase in total entropy.

Act III

In the meantime, Alice and Bob have gotten a divorce and are once again financially independent. They've lost their sense of family ties, but all three still love playing their game.

Bob: Let's see if we can develop a game that doesn't have this eternal drive to make everything conform, but also doesn't let everything get completely out of control.

Alice: We should also be able to play without going into debt.

Charles: Take biological evolution, for example. There the capable have all the luck, but the trees don't grow in heaven.

Alice: I once read that evolutionary equations aren't linear.

Bob: That's sounding terribly complicated. Please spare me anything more complicated than squares.

Alice: Thermodynamics can't describe these kinds of game rules. When one of us has all the money, then they'll stabilize in this state and the entropy will no longer grow.

Charles: But on the other hand, another property, let's call it fitness, will at least never decrease. This is what each person is willing to bid and play.

Bob: So let's say that we've got the following rules. Each person places a certain portion of his or her wealth into play. This speculation turns out to be a success, and the original amount gets doubled. After this success we go into the next round.

Alice: You mean that when I start out with V_a and speculate—let's say $1/4 V_a$—I'll have doubled my investment in the first round and have

$$(1 - 1/4)V_a + 2 \times 1/4 V_a = (1 + 1/4)V_a .$$

What you're calling my fitness is what I end up with as profit: $V_a/4$.

Bob: If that's the case, then I'll invest $V_b/3$.

Charles: And I'll go ahead and invest $V_c/2$.

Alice: Come on, guys; you can't do that. Remember, our top priority was that the total amount of money has to stay 1. Otherwise we just have meaningless inflation.

Bob: Then there'll be a skim taken off after each round, and the entire amount will be brought to normal.

Charles: Your investments determine the total fitness

$$F = 1/4 V_a + 1/3 V_b + 1/2 V_c .$$

And when I skim off the portion F from each individual income, then the sum balance should remain even.

Alice: How can you tell?

Charles: Because I'm so good at math. Our evolution rule E looks like this: when we invest a portion of our wealth, then we'll have $V(1 + s)$ going into the next round. In our case our account balances after being skimmed at the end of the first round look like this:

$$V_a\left(1 + 1/4\right) - V_a\left(1/4\,V_a + 1/3\,V_b + 1/2\,V_c\right),$$

$$V_b\left(1 + 1/3\right) - V_b\left(1/4\,V_a + 1/3\,V_b + 1/2\,V_c\right),$$

$$V_c\left(1 + 1/2\right) - V_c\left(1/4\,V_a + 1/3\,V_b + 1/2\,V_c\right).$$

When I add up the new account balances, I get

$$V_a + V_b + V_c + 1/4\,V_a + 1/3\,V_b + 1/2\,V_c - (V_a + V_b + V_c)\left(1/4\,V_a + 1/3\,V_b + 1/2\,V_c\right).$$

When the total amount of money is $V_a + V_b + V_c = 1$, then even after the first round,

$$1 + \left(1/4\,V_a + 1/3\,V_b + 1/2\,V_c\right)(1 - 1) = 1,$$

and in this way the total amount $= 1$ after each round, regardless of what the original distribution was.

Director's interruption: Bravo! You've just discovered the discretization of Fisher's evolution equation! Ronald Fisher used populations instead of players, and the V's were their sizes and not incomes. The process of evolution should produce natural selection, and he was thrilled to find that with his equation the average fitness always increased.

Alice: How do you know all that? Are you sure that we're not just getting into debt again?

Charles: Absolutely not. The total fitness is certainly less than 1, even $1/4\,V_a + 1/3\,V_b + 1/2\,V_c \leq 1/2$ for all V distributions equaling 1, so that the devaluation can never outweigh the income.

Bob: And despite your cleverness, you can still get the short end of the stick. If I got all the money, then it'd stay that way: $(0, 1, 0) \rightarrow (0, 1 + 1/2 - 1/2, 0) = (0, 1, 0)$.

Alice: But the same thing's true for me, too:

$$(1, 0, 0) \rightarrow \left(1 + 1/3 - 1/3, 0, 0\right) = (1, 0, 0).$$

Charles: That goes for me, too—this is where evolution gets stuck. In general, though, the total fitness increases. After the first round the total fitness

$$1/4 V_a\left(1 + 1/4\right) + 1/3 V_b\left(1 + 1/3\right) + 1/2 V_c\left(1 + 1.2\right) - \left(1/4 V_a + 1/3 V_b + 1/2 V_c\right)^2 \geq 1/4 V_a + 1/3 V_b + 1/2 V_c,$$

although the equality only holds when one V equals 1 and the other equals 0. It's only when one of us has all the money that the fitness level pans out.

Bob: This greedy kind of distribution when one person has everything and the others have nothing is luckily an exception to the rule.

Charles: Exactly. Otherwise the total fitness would always be increasing. We can see this clearly when we enter everything in my triangle like in Figure 6.1. That's where all points are trying to reach the distribution where I have it all (Figure 6.4).

Alice: This means that the income of one wealthy person must always try to approach the total amount of 1. You've thought up a doozy of a game.

Charles: Nothing ventured, nothing gained.

These games have shown that both rules that lead to disorder and rules that create order can be expressed mathematically. There are mechanisms that demand orderly structures, but the question is when do they take effect and when does the natural urge for chaos have the upper hand.

SECRETS OF WATER

*We see water as something mundane and don't have an inkling
how extraordinary it is.*

The compound H_2O has extraordinary properties that have made it our
elixir of life. The question arises as to where a secret could be hiding; every
H_2X compound has one X atom and two H atoms. The only difference
in structure can be the angle between the lines making the connection
between the X and the two H atoms hook up. For water, this is 104°, quite
obtuse, but this is actually what determines our welfare. We will now
explore exactly why this is.

By stretching out a water molecule only a little, it's possible to build one
of the simplest crystal forms, the tetrahedron structure. A tetrahedron consists of four identical equilateral triangles. It has four corner points connected by six edges. Each two connected edges form a 60° angle. The H_2O
molecules adapt so that the O atom is sitting in the middle of the tetrahedron and the two H atoms on different corners. In the crystalline phase,
as ice, each side is used as a side of the next tetrahedron, and each corner
point is part of six tetrahedrons. The sum of the angles is evened out: 6 ×
60° = 360°.

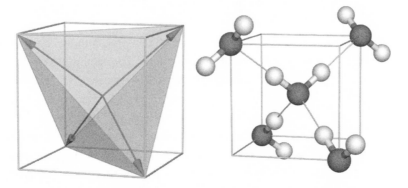

Figure 6.10: Diagram on left is of a tetrahedron inscribed in a cube.
Diagram on right is of the idealized tetrahedral structure of water.

The question is what happens to the angle between the two Hs and the O when the O is sitting in the middle of the tetrahedron. Naturally, we could calculate this angle on the computer, but this isn't a satisfactory course of action for math aesthetes. For them, it's the elegance of the thought process that's important; the journey is the destination. There's a very nice argument how we can painlessly solve this tricky geometry problem. Let's call the connecting lines between the middle of the tetrahedron and the corners v_1, v_2, v_3, and v_4 and then we can add up these objects in a normal way by linking them together. They are what's known as "vectors." The fact that the center is also the center of gravity is expressed by $v_1 + v_2 + v_3 + v_4 = 0$; they all balance each other out. There is also a multiplication between two vectors u, v, which is called the scalar product. It's written as $(u|v)$ and defines the length of the vertical projection from v to u times the length of u. It is symmetrical and distributive, meaning that $(u|v) = (v|u)$, and $(u|v + w) = (u|v) + (u|w)$. With this we've got the tool we need to solve our problem: the scalar product of a vector with itself is apparently the square of its length, so that $(v_1 + v_2 + v_3 + v_4 \mid v_1 + v_2 + v_3 + v_4) = 0$. Now let's use its distributiveness and multiply it out just like with common numbers. All four vectors are also equally good, particularly $(v_1 \mid v_1) = (v_2 \mid v_2)$, and so forth. All of them have the same length. Even the products of two different v's picked at random are the same. All of them have equal rights—$(v_1 \mid v_2) = (v_1 \mid v_3)$—and this holds true for every other pair. When we multiply the sums of all four vectors by the sums themselves, then we get a total of $4 \times 4 = 16$ summands. Four are the products of each v multiplied by itself; they are all the same, and we can choose this length as a standard unit of measurement. The remaining twelve of these products are from two different v's each, and are also all the same; we call them c (c stands for the cosine of the angle between two vectors, but the reader shouldn't be bothered with things like that). c defines the length of the projection of one v onto another and has a value between -1 and $+1$. Negative c means an obtuse angle, positive c an acute angle, and $c = 0$ means a right angle. $0 = 4 + 12c$ tells us that the sum of all v's has the length of zero, and in this way we get

$c = -1/3 = -0.3333\ldots$

The angle between two v's is therefore obtuse. For a water molecule, this projection actually equals $-0.241922.\ldots$. It doesn't have to be bent out of shape much to fit into a tetrahedron. With a molecule with an acute angle, we'd need an act of violence to make it fit. With a lot of H_2X molecules, the two H branches are almost perpendicular to another; the angle of H_2S is 92.2°, H_2Se is 91°, and H_2Te is 90°, so that c is almost always 0. It's therefore not so easy to build a tetrahedron out of any of these compounds.

But why are tetrahedrons so important to us? They give ice and water their life-friendly structure. The tetrahedron structure is pretty loose; it's possible to break it up with minor changes and melt ice into water. It's true that at small distances the structure of water isn't all that different from ice, and the tetrahedrons stay basically intact. Occasionally, through thermal motion, other water molecules will slip into the empty spaces, and the density in melted form is somewhat greater than in solid form. This is known to happen to water at up to 4° C, which is what we owe our ice-free bodies of water to. If this weren't the case, then ice would sink to the bottom, become thermally insulated, and wouldn't melt anymore. All the waters of the earth would thus eventually freeze from the bottom up, and all that would remain of our mighty oceans would be huge lumps of ice with a bit of thawed water on the surface. It's hard to say how life would have developed in such circumstances, but it would definitely have had a much more difficult time. What enormous consequences a tiny difference in degree can have!

The fact that the tetrahedral structure is kept on a small level also means that the melting and boiling points of water are much higher than those of comparable substances. The heat of evaporation of water is also higher than that of other H_2X compounds, which is a sign that the water molecules are bound together quite strongly. This is why water is so accessible on earth; it hasn't all evaporated like on Venus. On the other hand, it hasn't all frozen like on Mars. You could say that we don't need divine providence to explain this. If the situation were different, then life would emerge on another planet, or use other substances. It's difficult to say in theory how well that would work out. That's why the planned Mars landing is so exciting, as then we will see whether life could also emerge under completely different conditions.

The geometric anomaly of water has other, subtler consequences. Its specific heat—the energy that it stores when it's warmed up by one degree—is greater than that of any other liquid. In cal/g units it's 1.00.

To compare, glycerin has 0.57; benzene, 0.40; olive oil, 0.47. Due to the high specific heat, the oceans make ideal storage areas for heat, ensuring a stable climate. It has also been taken care of that oceans don't evaporate too quickly; the heat of evaporation for water is unusually high. In cal/g it's 538; to compare, for sulfur, it's 70; for ammonia, it's 327; for ethanol, it's 204; and for mercury, 68. We use the heat of evaporation as an effective means of cooling off; we sweat to release heat through evaporation. In this way, everyone can share in the magical powers of water. The pond skimmer enjoys its high surface tension, and a higher-than-average dielectric constant provides favorable conditions for high ionic density, which is in turn crucial for many biochemical mechanisms. We could go on and on singing water's praises, but we don't want to get repetitive.

Life is like a plant that curls upwards on the small anomalies. In this way it can create large anomalies, which provide life with new fuel. The oxygen in our atmosphere is an anomaly that was created by life, which then in turn made it possible for life to flourish on land and therefore made our very existence possible. In the next chapter I'll discuss a vision that intimates that the triumphal soaring of life will someday break out of the planetary framework and conquer the universe.

The Anthropic Principle—
Or Could You Patent the Universe?

In order for life to exist, an enormous number of coincidences must have come together during the evolution of the universe.

L ET'S TAKE another look at the powerful drama of the creation of the world:

A minuscule spark in an embryonic space filled with an ur-substance of dark energy with negative pressure was the beginning of it all. The repulsion it created swelled this tiny fragment of space, but the ur-substance regenerated itself everywhere so that more and more of this miracle stuff was formed. The repulsion grew beyond measure, and the expansion became an explosion of unimaginable force, the big bang. It shot open the gates to the underworld ever wider and let this matter come bursting forth. This matter developed into the particles we have today, all 10^{88} of them. It had energy and only positive pressure so that it took over the reins from the ur-substance and slowed down the explosion. However, there was still enough momentum to continue expanding for another 10^{10} years. We are left with just a pale reflection of the beauty of the underworld; its symmetry was crushed and destroyed in the big bang. The unified force was split into four forces, which continue to become more and more alienated from one another. The masses of the elementary particles appear completely disorganized like a scrapheap. Perhaps even our three-dimensional space is only a scrap from an ur-space of greater dimension and symmetry. Most of the structures created kept on burning so that space was dominated by a fine layer of photon and neutrino ash. Yet there still remained something else in this background which would be the progenitor of all chemical elements: atomic nuclei—75 percent hydrogen, 25 percent helium lightly seasoned with some heavier atomic nuclei and the electric charge of all nuclei neutralized by electrons. The miracle had occurred. These apparently chaotic bits and pieces joined together to become the foundation of matter.

All of this happened within three minutes. It was to take much longer for the preparation of matter to be completed. Billions of years, to be precise. But the big bang was explosive enough so that we won't be pressed for time and can avoid having everything collapse before we're through.

As the universe continues to expand and cool down, gravitation starts its work at a leisurely pace. It forms gas clouds into balls. The larger this kind of condensation nucleus is, the greater the amount of matter that falls into it. Initially, the thermal pressure prevents a further concentration so that the sirocco continues to swell until finally the temperature reaches 20 million degrees in the innermost core. Millions of years pass until the reaction $P + P \rightarrow D + e^+ + v$ takes place and a star is born. But there's no one around to enjoy the sunshine. The world consists of just hydrogen and helium, and that's too meager for life.

From now on, the flame is kept on low. No more energy is emitted than can be produced in the universe's nuclear power plant. In this way, the energy budget is kept balanced for billions of years. Everything must come to an end, and when this source of energy is used up, the game with negative specific heat begins again. The star then survives on its gravitational energy reserves; if it wants to shine, it has to compress and heat up until a heavier flammable fuel ignites. In this way the materials that will one day support life get baked in the cosmic oven—carbon, nitrogen, oxygen, down to iron. After billions of years, dinner is ready. The question is who opens up the oven. The Lord has taken care of that, too. The star opens up by itself; it bursts and in a gigantic supernova the cosmic fertilizer is flung into space.

Then the game begins all over again: cosmic gas clouds accummulate to form stars, which this time around are pregnant with heavy elements. These fall away during the condensation process and build smaller planets. Somewhere there's also one that has got everything just right: a stable orbit, climate, and temperature, so that water can go from solid to liquid to gas and back again: a sufficient amount of water and other gases, but not too much to prevent the greenhouse effect from burning everything. Over the next billion years, a wondrous system cautiously emerges: the first living cells. These prove to be too egoistical and are only concerned with self-improvement to the point of perfection. They spend the next three billion years in a sterile dead-end.

At that point, finally, a few recognize that it's sometimes better to cooperate than to compete. They surrender part of their autonomy and, in doing so, become more efficient. These are the multicellular organisms. The new system catches on like wildfire over the entire earth. New species and varieties are constantly being created. Under the pressure of competition, they constantly become larger, faster, and stronger, until finally gravity puts an end to development in this direction. The system therefore freezes for a few hundred million years. But the dinosaurs did not fulfill the Lord's intention, and he threw an iridium asteroid to the earth sixty-five million years ago and destroyed these upstarts' basis of existence. Everything started all over again. Smaller animals were forced out of the jungle into the savannah due to the climate change, and the harder fight for survival honed their senses so that they developed what the Lord intended to carve out of them: a mind. They are able to recognize his laws and consider themselves to be made in his image: the human being.

Who can be left cold by this fabulous tale? Still, what does this teach us?

The lesson to be learned is called the anthropic principle. This states that, at a crossroads in development of the universe, the turn has always been taken that would ultimately allow the creation of human existence. This has been illustrated for us in numerous ways and could be jokingly described as if someone wanted to patent the universe; they'd be turned away immediately—the reason being that such a complicated mechanism, which only functions when so many random events exactly coincide, would always fail in reality.

There are three degrees of the anthropic principle:
1. The weak anthropic principle:
   We see everything in the universe as having the constraint of allowing human existence.
2. The strong anthropic principle:
   The universe must have the constraint of allowing human existence.

3. The final (or eschatological) anthropic principle:
The laws of nature are construed in such a way that human beings
will ultimately be able to populate the entire universe.

Remarks:

re 1. Practically a tautology, as if the universe weren't conducive to life
in our environment, we wouldn't be able to observe it. Although necessarily true, even this weak form is occasionally met with skepticism.

re 2. What does "must" mean? Who would punish it for disobeying?
Does "must" really mean "wants to"?

re 3. A grand vision, even though it seems utopian. According to it, in
the cosmic evolution the human mind is more than a tiny spark that immediately fades away—rather it ignites a fire that can ultimately consume the
entire universe. The laws of physics really mean the laws of nature. The
question is whether they draw the limits of human development, or
whether psychological factors like selfishness, irrational fears, delusory
beliefs, and so forth determine our fate for billions of years.

There are different ways to interpret the anthropic principle.

On the one hand you can be Darwinist and say that there were innumerable sparks like the one that ignited the big bang. That way there
would be many worlds, and in one of them everything just happened to
fit and that one was ours. This is logically possible, but there's no trace of
any of these other worlds. In science, objects that are not subject to observation are considered to be ideological ballast and are eventually thrown
overboard. That's why many reject this explanation. Perhaps it would be
wise to give this suggestion a chance; from a purely logical point of view
it is acceptable, after all. If there were so many different big bangs, there
might be a few more small bangs in our universe, aftershocks of the big
big bang. In fact, it's possible to see phenomena billions of light years away
with energies that eclipse a supernova thousands of times. These could
be possible candidates for the small bangs.

On the other hand, it seems possible that the principle of self-organi-
zation is as general as that of increased entropy, and higher organized
beings could have come into existence under completely different circumstances that appear to us to be hostile to life. They could, for example,
exist on the surface of a cooled-down neutron star, but would then have
to be microscopically small. The reason I don't believe in their existence
is that the orbital period of particles in the nucleus is 10^{-22} s, 10 powers

of 10 shorter than that of normal matter. Biological evolution would have to be correspondingly faster in its development and last about a month rather than 10^9 years. Because there were supernovas already taking place billions of years ago, there must be neutron stars just as old. If there were civilizations on them, then they must have already been able to reach the apex of their possibilities, their Ω point. They should have also developed a need to communicate, but we've never heard a thing.

From a purely rational standpoint, we can't do all that much with the anthropic principle. We also can't make it a law of nature; it's too imprecise and it can't be mathematized. Nonetheless, it is responsible for several scientific breakthroughs. For example, using the weak anthropic principle, Fred Hoyle was able to predict certain properties of the atomic nucleus of ^{12}C based on the fact that this was the only way the synthesis of other heavy atomic nuclei could function in stars. Without a more exact formulation, we can be sure only of one thing: the universe appears to follow the urge to create increasingly complex structures. What do we mean by "urge"? Nature is inanimate and doesn't feel urges or feelings, but follows laws dictated by a higher power. Human beings have always been aware of this kind of higher power or powers and had a wide range of ideas about them. I would also like to speak of God, but without trying to get to the bottom of his nature. I wouldn't dare to try and force him under the yoke of my human logic and pinpoint him dogmatically. After these preliminaries, I will try to summarize our observations from our forays into the universe in a theistic version of the anthropic principle.

God directs the development of his creation so that his image, human beings, can come into being.

"Directs" means in a way that can appear to be random or necessary. Whether we are the only images of him—in actual fact, the pinnacle of creation—is something we don't know for sure. But maybe that's not all that important anyway. In any case, we have a privileged position in the universe. Privileges also entail responsibilities, and the knowledge given to us not only brings us understanding, but also a sense of ethical behavior. After we have recognized the will of God, we must use this to guide our actions. I don't want to go so far to consider if we will fulfill the prophecy in the eschatological anthropic principle in a few billion years, and interpret the commands in the Bible to say that we should dominate and rule over the entire universe. We have more pressing goals closer to home, and it would already be a great blessing to have the Ω point from Teilhard de Chardin in 40 million years. Perhaps our insights into the nature of creation can help us achieve this goal.

Turning back to our original question, naively put as: Can we explain the evolution of the universe naturally, or do we need supernatural intervention?

Our panorama of evolution has always shown mathematizable laws, which as such determine the future through the present. But even the present is known only to a finite degree, meaning that there is also a measure of uncertainty in the future. These laws also contain parameters that are unknown from the start, so the location of this uncertainty is not fixed by basic laws. Our universe is a sensitive structure, and there's a thin line between success and failure. The final product, the current universe, seems to us to be random, even extremely improbable, because it's only when these parameters have certain exceedingly rare values that everything comes together so harmoniously. Those are the facts. They are interpreted in different ways; this is where there are disagreements. Let's try to formulate three approaches:

1. Many things will become better understood over the course of the development of humanity. Maybe we will someday have a "theory of everything" to explain it all. Right now we've got to accept what's not understood as fact. There's no point in asking why.

2. It now looks as if the universe is infinite in size. This means that all possible values of the fundamental physical constants and initial conditions could be realized in different parts of the universe. We are living in an area conducive to life. To explain why, we just need to point to the weak anthropic principle, which is practically a tautology.

3. The laws of nature have come about through an evolutionary process that necessarily led to the present life-friendly form of the universe. Lee Smolin suggested the following mechanism as an explanation: perhaps the singularity after the gravitational collapse is not a final death, but in quantum gravitation this becomes a new big bang and a new world can prosper. This world has inherited from its earlier existence the values for the fundamental physical constants, with just a few small errors. Each black hole therefore has descendants whose constants allow the creation of black holes. This is how these kinds of worlds are reproduced. A universe with

completely different constants wouldn't even contain stars, much less black holes, and would therefore remain sterile. This would create a supercosmos with an infinite number of worlds, but those that can generate black holes would dominate. These worlds happen to be those in which life is also possible. According to this view we're still a product of chance, but also a very probable, almost necessary result of cosmic evolution.

It's clear that positions 2 and 3 are quite speculative, rather than predictable, as they would need to be to become a fixed part of science. As the basis for an entire *weltanschauung* they're too shaky, but they're tenable as possibilities for consideration. We are simply not able to give a clear answer to our first question, if we can provide a natural explanation for cosmic evolution; this has proven to be beyond us. We're able to explain a lot of things that had previously appeared incomprehensible, but only by introducing new, strange, and wondrous explanations. Our efforts were not in vain. We now have a much more magnificent picture of cosmic evolution than we did even a few decades ago. Our journey was our destination. It led us by many miracles of nature and taught us to be in awe of the incredibly fine-tuned blueprints of the universe. It opened up our eyes to the responsibility of humanity, filled with cosmic relevance. As the masters of the earth, we have been handed the flame of life; are we going to let it fade away, or will it shine over the entire universe?

This is where the human mind in its entirety, where science and religion are called upon.

Isaac Newton

im zwölften Lebensjahre

Explanation of Symbols and Glossary

EXPLANATION OF SYMBOLS

Symbol	Meaning
$a = b$	a equals b
$a \sim b$	a roughly equals b
$a > b$	a is greater than b
$a < b$	a is lesser than b
$a \times b$	a times b
$a : b$ and a/b	a divided by b
a^2	a times a, a squared

GLOSSARY

Big bang
It appears that the cosmos is the result of an extremely powerful explosion that ripped apart the embryos of all matter 15 billion years ago. This explosion is called the big bang.

Bosons and fermions
There are two different kinds of elementary particles. These have completely different properties and are named after the people who discovered their behavior. Distinguishing between them provides the crucial key in understanding the structure of matter.

Chaos and ergodicity

The intuitive feeling of chaos can be specified for a dynamic system. The dynamic consists of the specifications of trajectories in the state space, with exactly one trajectory intersecting each point. Such a system is called *ergodic* when the trajectories originating in the smallest areas of the space evenly cover the entire space in a time average. The system is *chaotic* (or mixing) when this will happen just as well not only in a time average, but given sufficient time.

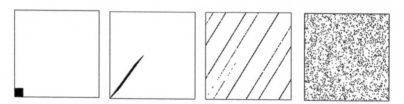

Figure A1: The simplest example of a mixing dynamic that is expanding in one direction and compressing in the perpendicular direction. The space is a periodic square; when something exits stage right, it can enter stage left. The left-hand illustration corresponds to the initial state (t=0). The other illustrations depict from left to right the state after one, four, and eight iterations.

Compton wavelengths and Planck's units

The speed of light c assigns a frequency f to a length L as follows:

$c = Lf$ or $f = c/L$ or $L = c/f$.

The quantum h then assigns to a frequency f, an energy amount such that a light quantum of this frequency has an energy E:

$E = hf$ or $f = E/h$.

If E is the rest energy of a mass m, $E = mc^2$, the mass m has a length λ, the Compton wavelength,

$\lambda = c/(mc^2/h) = h/mc$.

The significance of λ is that a light quantum with this wavelength has sufficient energy to produce a particle with a mass m. Planck's units are arrived at not by arbitrarily picking a particle mass, but rather they are defined through the gravitation of the universal Planck's mass M_P. This is achieved through the requirement that the gravitational energy of two Planck's masses with the distance of their Compton wavelengths be the negative of their rest energy.

Expressed in equations:

$$GM^2{}_P/(h/M_Pc) = M_Pc^2 \text{ or } GM^2{}_P = hc \text{ or } M_P = (hc/G)^{1/2}.$$

Planck's length is this Compton wavelength, and Planck's time the time needed for light to cross it.

Elementary and fundamental particles

There were originally three of them: the electron (e), which populates the atomic cloud; the proton (P), the nucleus of the hydrogen atom; and the photon, the light quantum (γ). But then the proton got an electrically neutral little sister, the neutron (N), and the electron a positive mirror image, the positron (e^+). Then things with a short lifespan and masses somewhere between electrons and protons appeared; these were called mesons. The faster the observations could be made, the more elementary particles were found, until there were soon more than a hundred. Then the Standard Model thinned out this jungle and replaced them all with three types of particles: the quarks, the leptons, and the gauge bosons. As the term "elementary particle" was already in use, these elementary building blocks were called "fundamental particles."

Foucault's pendulum

This pendulum is hung so that it can oscillate in every plane without friction. Through the earth's rotation the plane also rotates for the observer on the earth. If the pendulum were to stand on the North Pole, then its oscillation plane would be stationary in the reference system in which the earth rotates; seen from the earth, the oscillation plane makes a 360° rotation in twenty-four hours.

There seems to be a force acting on the pendulum as seen from the earth; this is called Coriolis force.

Gamow factor

It costs infinite energy to set two positive charges on the same point. As quantum mechanics is with finite energy, this is extremely unlikely to happen. The Gamow factor says exactly how likely this is.

Red giants

When a star has burned its hydrogen into helium, its interior becomes denser and hotter until the temperature is reached where the helium can be transformed into carbon and oxygen by nuclear fusion. The gravitational energy won by this contraction swells up the exterior of the star, which then cools off to some degree. The star then becomes gigantic, cooler on the outside and therefore redder. This explains the name "red giant." This process takes millions of years and is not a cosmic catastrophe.

Torus

A mathematical construction that generalizes a circle onto several dimensions. In two dimensions this is realized with the surface of a "doughnut" (periodic square). The principle behind it is that whenever something goes too far in one direction, it can get back in from the opposite side.

Powers of Ten

NUMBER	WRITTEN AS	DECIMAL POWER
Ten	= 10	= 10^1
Hundred	= 100	= 10^2
Thousand	= 1,000	= 10^3, etc.

These are just different ways of saying the same thing, but to save paper we'll use the decimal power, as it's much shorter to write 10^{17} than 100,000,000,000,000,000, and there's also no good name for that either. The prefixes femto = 10^{-15}, pico = 10^{-12}, nano = 10^{-9}, micro = 10^{-6}, milli = 10^{-3}, kilo = 10^3, mega = 10^6, giga = 10^9, tera = 10^{12}, peta = 10^{15} are often used, but there's nothing comparable in everyday language ("ur", "super," and their comparatives can't be expressed numerically). I'd like to save the reader the trouble of these strange prefixes and express everything using powers of ten. Using them, multiplication is a matter of simply adding the exponents together.

$10 \times 10 = 100$, so $10^1 \times 10^1 = 10^2$

$10 \times 100 = 1,000$, so $10^1 \times 10^2 = 10^3$, generally

$10^n \times 10^m = 10^{n+m}$.

n or m can even be negative.

$100 \times 1/10 = 10$, so $10^2 \times 10^{-1} = 10^1$.

n and m don't even have to be whole numbers; we'll even be using $10^{0.5}$. This number multiplied by itself should produce 10; it is therefore somewhat larger than 3, as $3 \times 3 = 9$ is already almost 10. That's enough of that. What's closer to 1 is rounded down to 10^0 (which is equal to 1) and what's closer to 10 is rounded up to 10^1.

APPENDIX 3

Lord Kelvin's Estimate of the Sun's Age
Using Modern Terminology

*The thought process is simple—when gravity compresses gas clouds
into balls, it gives them a speed and generates thermal energy.
This in turn produces a photon gas that escapes the sun's surface
as sunshine. When sunshine has taken all the heat away, the sun will
be extinguished.*

FIRST WE NEED to find out the gravitational energy of the sun. This
has come about as gas clouds have crashed into the sun from great distances and therefore gained speed. Their kinetic energy is equal to their
(negative) gravitational energy, as the sum of the two can't change. If a
body crashes into the sun from a great distance, then near the earth's orbit
it has a somewhat greater speed than the speed of the earth in its orbit
around the sun; which is 30 km/s. We don't feel a thing from this race
through space; it's the result of:

Intermediate calculation: according to the definition

speed = (distance covered)/(time needed).

The earth's orbit around the sun has a length of 10^9 km and the year is
$10^{7.5}$ s, so that we have an orbital speed of 10^9 km / $10^{7.5}$ s = $10^{1.5}$ km/s ~
30 km/s. If the body falls farther all the way to the sun's surface, its speed
increases tenfold. Why is that?

Intermediate calculation: the potential gravitational energy is proportional to

1/(distance to the sun).

Because the sun's diameter is 1/100 of the distance between the earth and the sun (the reader may try and cover the sun with outstretched arm [~1 m] and with a thumb [1 cm thick]), in falling farther past the earth to the sun's surface, the distance to the sun decreases a hundredfold, which means that the potential energy increases by 100. The kinetic energy must increase by the same amount. It goes along with (speed)^2, so that the speed increases tenfold to about 300 km/s. This means we'll land on the sun with a speed of 300 km/s. (Using more precise calculations it's even 620 km/s.)

This corresponds to a dreadful heat, as at room temperature (= 300K) the thermal velocity v of the molecules ~1 km/s. Like the kinetic energy, the temperature goes with (speed)^2, so that in appropriate units $v^2 = 300$ K. With three hundred times the speed that our test object has reached falling into the sun, it's also generated a temperature of $300^2 \times 300$ K ~ 10^7 K, 10 million degrees. The reader might object that the sun's surface only reaches 6,000 K, but this is because it has been cooled off by sunshine. It's only on the inside where you can still find the horribly high temperatures that accumulated as the sun was formed into a ball. We can't measure them directly, because our neutrino eyes can penetrate into the sun's center. They are sensitive to temperature, so that today we know for sure that it's about 15 million degrees in there.

We've still got to estimate how long the sunshine needs to remove this amount of heat. In keeping with quantum theory, we should think of sunshine as a gas of light quanta (= photons), with each photon having the amount of energy a particle would at 6,000° (= temperature of the sun's surface). The density of this gas is one photon per wavelength cubed. The median wavelength of sunlight is 10^{-6} m, meaning that the volume of the cube is $(10^{-6}$ m$)^3$; therefore, the photon density on the sun's surface is one photon per $(10^{-6}$m$)^3$. They escape from here with the speed of light (300,000 km/s). This says it all, because this density is 10^{-12} times the median density of matter in the sun. The energy density of the photons on the surface is further reduced by (surface temperature/interior temperature) = 10^{-3} compared to the energy density of matter in the interior. It is $10^{-3} \times 10^{-12} = 10^{-15}$ times smaller than the thermal energy density in the sun's interior. This is what the sunshine has to transport. If it were to have the same energy density, it would take a layer as thick as a light second each second. But it is weaker by 10^{-15} and can do this only at a fraction of 10^{-15}. In 10^{-3} s light travels a little less than 1/10 mm, which is the thickness of the layer of the sun's surface from which the energy escapes through sunshine per second into the universe. We know the sun's radius = 100 × earth's radius ~ 10^6 km = 10^{15} × 10^{-3} mm. The conclusion reached by these calculations is that the lifespan of the sun is 10^{15}s, which roughly equals 30 million years, which proves that there was something missing in Lord Kelvin's assumptions. We've used the photon, but the quantum structure of light is irrelevant here, and Lord Kelvin couldn't achieve any other result even without it.

How Much Does the Universe Weigh?

The answer is: nothing. The negative gravitational energy offsets the positive rest energy of the matter so that as a whole the cosmos has zero energy, and therefore zero mass null. To prove that gravitational energy = rest energy, we'll show that a body falling into the universe travels at the speed of light due to gravitation.

THE IDEA behind the following estimation is quite simple. We start with a body whose velocity of fall v we've already calculated. Then we'll see how much its M/R differs from the M/R of the universe. The formula that determines v, $v^2 = GM/R$ then tells us how the v of the universe is different from the v of the body. We'll use the sun as a reference, knowing that the sun's radius is 10^6 km. For the entire universe we have radius = speed of light times the age of the universe,

$R = 3.10^5$ km/s $\times 10^{10}$ years $= 10^{5.5}$ km/s $\times 10^{17.5}$ s $= 10^{23}$ km $= 10^{17}$ sun's radius.

We had calculated that the fall velocity to the sun v_S is 300 km/s. V^2_S will serve as a benchmark; for the entire universe we need to multiply this with the proportional relationship of the masses and divide it by that of the radii. Therefore, the following relation holds true for v_U, the fall velocity into the universe:

$v^2_U = v^2_S (M_U/M_S) (R_S/R_U) = v^2_S 10^{22} 10^{-17} = v^2_S 10^5,$

where R_S represents the sun's radius and R_U the universe's. The same convention is true for the masses M. We're also using the figure that there are 10^{22} stars in the universe and that $R_U/R_S = 10^{17}$. To arrive at v_U, we still have to find the square root, simply by cutting the exponents in half:

$$v_U = 10^{2.5}\, v_S = 10^{2.5} \times 10^{2.5} \text{ km/s} = 10^5 \text{ km/s} \sim c,$$

almost too fast for the cosmic speed police.

Antigravity at Work

In Einstein's theory, gravity can also become repulsive, which means that everything gets repelled in all directions.

THE ONLY WAY to prevent the total collapse is when $E + 3p$ is negative; then the implosion becomes an explosion. According to Friedmann, the following is true for the world's radius $R(t)$ in Einstein's theory:

$$R''(t) = -R(t)(E + 3p) .$$

We've thrown the unnecessary ballast of constants overboard and replaced them with suitable units; R'' is acceleration, the second temporal derivation of $R(t)$.

There's no reason to be sad if you, dear reader, are inexperienced in differentiations. For our purposes it's sufficient when we think of the derivation as a change in the time unit. The latter is Planck's time, and is infinitesimally small for all practical purposes. The expansion speed, the derivation of $R(t)$, is then $R'(t) = R(t + 1/2) - R(t - 1/2)$, and its derivation, acceleration

$$R''(t) = R'(t +1/2) - R'(t -1/2) = R(t + 1) + R(t - 1) - 2R(t).$$

According to Friedmann then

$$R(t + 1) + R(t - 1) - 2R(t) = -R(t)(E + 3p)$$

must hold true for all times t. When E and p are not dependent on the time t, then this is carried out by $R(t) = 10^t R(0)$. By inserting this into Friedmann's equation we get

$$10R(t) + (1/10)R(t) - 2R(t) = -R(t)(E + 3p),$$

where $R(0)$ cancels itself out in the insertion process. This equation holds true for every $R(0)$, as long as

$$10 + 1/10 - 2 = -(E + 3p)$$

is true for the values we've chosen for E and p. As $10 + 1/10 - 2 > 0$, $E + 3p$ has to be negative; the numeric value can be corrected by changing the temporal scale when it isn't completely correct.

APPENDIX 6

A Game of Marbles

If there are only two possibilities for a phenomenon, it's easy to find out how chance will decide in N independent phenomena.

$\mathbf{F}$OR THOSE who are only interested in philosophical content, this appendix can be skipped in good conscience. For those of you who are curious to see how an expert goes about solving these kinds of problems, this section can perhaps provide some satisfaction. z_n represents the numerical registration of the distribution of black and white among N marbles. The formula for z_n contains a jumble of factors that first need to be cleared away with an appropriate notation. I'll call the product $1 \times 2 \times 3 \times \ldots \times N$, N factorial, and denote it by $N!$. To get used to this way of writing, let's start small. $1! = 1, 2! = 2, 3! = 6, 4! = 24. \ldots$

As $N! = N \times (N - 1)!$ it's only necessary to multiply the last result with the next number. If I go ahead and define $0! = 1$, then our proposition can be written as

$$z_n = N!/n!(N - n)!,$$

where n can be every integer $0, 1, 2, \ldots, N$. To see if this formula is correct, I'm going to replace the colors with two properties written with the signs "+" and "-". z_n is therefore the number of the different orders of N signs, where n times "+" and $N\text{-}n$ times "−" appear. Let's start small again.

$n = 0$: there's only "-", so there's only one order -, -, -, . . . -. In fact, $Z_0 = N!/0! \times N! = 1$.

$n = 1$: we have to incorporate a "+", and this happens in N ways:

+, -, -, . . . , -; -, +, . . . , -; -, . . . +, . . . ,-; -, -, -, . . . ,+.

The formula $z_1 = N!/1! \times (N - 1)! = N$ still holds true.

$n = 2$: here something new happens. You'd think that for the first "+" there'd be N spaces and then for the second $N - 1$, so that there would be altogether $N \times (N - 1)$ possibilities. But that means that we've counted everything twice, because you can't tell in a distribution which is the first "+" and which is the second. The order +, +, -, . . . , -, for example, appears twice in our count when I set the first "+" on the first space and then set it on the second. So I've still got to divide by two so that there are $N \times (N - 1)/2$ different orders. And in reality our magic formula says

$$z_2 = N!/2!(N - 2)! = N(N - 1)(N - 2)!/2!(N - 2)! = N(N - 1)/2,$$

when I use the definition of factorials and leave out the eternal × in multiplying.

We can already recognize the general law of formation. For $n = 3$, for example, there are N, $N - 1$, and $N - 2$ positions, respectively, for the first, second, and third +, but changing the three +s around won't lead to anything new. As there are $3! = 6$ possible alternative placements for three objects, $z_3 = N(N-1)(N-2)/6$, and that is once again $N!/3!(N-3)!$. In this way we get

$$z_n = N(N - 1)(N - 2) \ldots (N - n + 1)/N! = N!/n!(N - n)!$$

for every possible n.

To carve out the form used in our text, we need to tackle the factorials. We'll try it once as if all factors were to equal N in $N!$ and take the liberty to write brazenly $N! = N^N$. This is a hopeless overestimation of $N!$; a better expression is $N! = (N/e)^N$. $e = 2.7 \ldots$ is one of the magical numbers in mathematics, but we don't need it as in z_n it cancels itself out. We now arrive at $z_n \sim N^N n^{-n}(N-n)^{n-N}$, but that still doesn't really bring much joy. So let's see if the formula can thin out near the disorder and set $n = N(1 + d)/2$, where d should be much less than one. Then all N^N cancel themselves out and we're left with

$$z_d \sim (1 + d)^{-N(1 + d)/2}(1 - d)^{-N(1 - d)/2}2^N.$$

This is still not ideal, but now's the time to keep in mind that $z_d = z_{-d}$, black and white are equal. z_d can only contain the even powers d^2, d^4, ..., and for small d can only be d^2. I still need a magic formula valid for N

$$(1 \pm d)^N \sim 2^{\pm Ndc},$$

where c nearly equals 1 to arrive finally at the desired form

$$z_d \sim 2^{cdN(1 - d)/2 - cdN(1 + d)/2} w_0 = 2^{-Nd^2c} w_0.$$

The Neutrino Rain of a Supernova

The neutrinos of a supernova build a bubble that expands at the speed of light.

L ET'S SEE what we'd feel from the neutrino flood if there was a supernova in our Milky Way about 1,000 light years away. By the time the neutrino flood has sped to us at the speed of light, it has built the surface of a ball with a radius of 1,000 light years =

$$(10^3 \text{ years}) \times (10^{7.5} \text{ s/year}) \times (10^{10.5} \text{ cm/s}) = 10^{21} \text{ cm radius.}$$

The surface of a ball is $4\pi(\text{radius})^2$, meaning that this surface is

$$10 \times 10^{21} \times 10^{21} \text{ cm}^2 = 10^{43} \text{ cm}^2 \text{ (10 stands for } 4\pi),$$

and inside are all of the neutrinos received for 10^{57} of the star's protons. One neutrino per proton means that in a few seconds $10^{57-43} = 10^{14}$ neutrinos per cm^2, 1,000 times the dose of the sun's neutrinos of 10^{11}.

Taking Newton's Vision Further

We are taking Newton's extrapolation from an apple to the moon further, from the sun to the center of the Milky Way.

NEWTON'S SUCCESS was such a smash that we'd like to test if the same force ties the earth to its orbit around the sun, too. Our goal is to see if the centrifugal force balances out gravity, if our condition:

gravitational acceleration due to the sun = acceleration through orbiting the sun

is met. I say "our" on purpose, as according to Newton, the motion is not dependent on the earth's mass. If this acceleration caused by the sun applies to us, then it applies to everyone. Even without the earth, we would orbit the sun in a year. If this weren't the case, then we'd lose astronauts to their own orbits while we keep traveling on ours around the sun. To test the equation above, like a schoolchild in the last row, we'll just want to copy our reflections on the moon's orbit. We'll check if the gravitational acceleration in the earth's orbit around the sun compared to the moon's orbit around the earth changes by the same factor as the centrifugal force. The right side is

distance / (orbital period)^2,

and the left

mass / (distance)^2.

On the right side there's a change of the ratio of the distances times the square of the ratios of the orbital periods in orbital acceleration—that is, of

(distance earth-sun/distance earth-moon) × (1 year/month)^2 = 400/144 ~ 2.8.

On the left side the gravitational acceleration changes by

(sun's mass/earth's mass) / (distance earth-sun/distance earth-moon)^2 = $10^{5.5}/(400)^2$ ~ 2.

2 instead of 2.8 is not bad for a first estimation when the huge discrepancies in masses and distances are taken into consideration. This spurs us on to be bolder and leave Newton far behind. We're going to go to the very end of our Milky Way, which Newton couldn't have foreseen. We're living in a spiral cloud with a diameter of 10^5 light years, and our spiral arm orbits the center in approximately 100 million years. The centrifugal force of the rotation needs to balance out the gravitational attraction here as well. We simply need to compare by how much smaller both forces in the Milky Way are to our orbiting the sun. If the gravitation is the same as in our solar system even in this cosmic scale, then both factors must agree.

The left side:

(distance spiral arm-center/distance earth-sun) × (one year/10^8 years)^2

= ($10^{4.5 + 7.5}/10^3$ in light seconds) × $(1/10^8)^2$ = $10^9 × 10^{-16}$ = 10^{-7}.
The right side:

(Milky Way mass/sun's mass) / (distance spiral arm-center/distance earth-sun)^2 = $10^{11}/(10^{9.5})^2$ = 10^{-8}.

So they're not exactly the same. The gravitation is a little too weak, but we also used very rough calculations. When you calculate more exactly, it comes out all right, you'd think, wouldn't you? That's not the case. The more precise the calculations, the greater the discrepancy becomes. We're forced to conclude that most mass is not visible, but comes from a "dark matter." There are several possible sources for this: extinguished stars, black holes, still unknown particles, but none of them is really conclusive.

Wise Guy's Homework Assignment

If you cross a planetary orbit, you can be flung out of the solar system!

I F A BODY FALLS to a planet from a great distance, its orbit around the planet forms a hyperbola, like a comet's orbit around the sun. In decelerating, we need to try and make sure that the direction of the hyperbola's axis jibes with the direction of Mercury's orbit. Then it will be pulled along by Mercury and finally reaches twice the speed of Mercury in this direction. This is shown in the following: we'll break down the velocity v_S of the satellite in both directions parallel and perpendicular to Mercury's orbit in the universe. Both components were $(0, v)$ at impact on the sun. The satellite therefore has the velocity component $v_S = (-v_M, v)$ in the Mercury system; through the collision the first changes sign but the second remains intact. Thus the veocity becomes $v_S = (+v_M, v)$, which is then $v_S = (2v_M, v)$ in the universe system. This is sketched out in Figure A9.

The satellite is pulled behind Mercury, so that attraction appears as reflection. The collision with Mercury has the same effect as a reflection on a moving wall, even though gravity is an attracting force. Everything here makes a mockery of common sense. The speed in the direction toward the sun remains, but due to the speed in the direction of Mercury's orbit, I'll actually end up passing the sun. I can even stay away from the sun altogether! After its Mercury adventure, the satellite's total energy E is, according to Pythagoras, half of the sum of the squares of the velocity components and the negative gravitational energy—that is,

$$E = 1/2(4v^2{}_M + v^2) - GM/R_M.$$

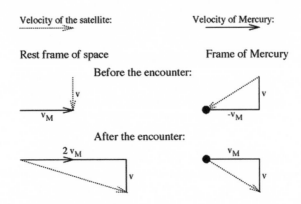

Figure A9: Meeting of a satellite and Mercury as seen from the system in which the universe is at rest (universe system) and the system in which Mercury is at rest (Mercury system).

The satellite gains the velocity v through the difference between the gravitational energies from the sun to the earth and to Mercury, and can be expressed with the help of the virial theorem through the planetary velocities as:

$$v^2/2 = GM/R_M - GM/R_E = v^2_M - v^2_E.$$

Were I once again to replace MG/R_M with v^2_M and add up everything, I would have

$$E = 2v^2_M - v^2_E.$$

At the same time, this is also the kinetic energy $v^2_\infty/2$ at the escape. Now the proportion between squared velocities of Mercury and the earth is exactly the inverse of the proportion of their distance to the sun, which is easy to measure. If I use this empirical value for $v^2_E/v^2_M = R_E/R_M$, and set E as $v^2_\infty/2$, I'll find that

$$v_\infty = 2v_M[1 - v^2_E/2v^2_M]^{1/2} = 1.9v_M.$$

That's around $5v_E$ as opposed to the professor's measly $1.4v_E$. It's therefore much better to get involved with Mercury first rather than to try and run away from the sun directly!

Index